Amin Nozariasbmarz

Thermoelectric Materials for Wearable Electronics

Also of interest

Chemical Energy Storage
Robert Schlögl, 2022
ISBN 978-3-11-060843-4
e-ISBN (PDF) 978-3-11-060845-8

Organic and Hybrid Solar Cells
Lukas Schmidt-Mende, Stefan Kraner, Azhar Fakharuddin, 2022
ISBN 978-3-11-073692-2
e-ISBN (PDF) 978-3-11-073693-9

Thermoelectric Materials.
Principles and Concepts for Enhanced Properties
Ken Kurosaki, Yoshiki Takagiwa and Xun Shi, 2020
ISBN 978-3-11-059648-9
e-ISBN (PDF) 978-3-11-059652

Nanocarbon-Inorganic Hybrids.
Next Generation Composites for Sustainable Energy Applications
Dominik Eder, Robert Schlögl, 2014
ISBN 978-3-11-026971-0
e-ISBN (PDF) 978-3-11-026986-4

Amin Nozariasbmarz

Thermoelectric Materials for Wearable Electronics

Body Heat Harvesting

DE GRUYTER

Author
Dr Amin Nozariasbmarz
Department of Mechanical Engineering
Rowan University
Glassboro, NJ 08028
USA
nozari@rowan.edu

ISBN 978-1-5015-2105-8
ISBN 978-1-5015-1481-4 (PDF)
ISBN 978-1-5015-1483-8 (EPUB)
DOI https://doi.org/10.1515/9781501514814

Library of Congress Control Number: 2026940484

Bibliographic information published by the Deutsche Nationalbibliothek
The Deutsche Nationalbibliothek lists this publication in the Deutsche Nationalbibliografie;
detailed bibliographic data are available on the internet at http://dnb.dnb.de.

De Gruyter and Walter de Gruyter GmbH are part of De Gruyter Brill.
www.degruyterbrill.com

Questions about General Product Safety Regulation:
productsafety@degruyterbrill.com

Cover illustration: Ryan Penn
Printing: CPI books GmbH, Leck

Contents

Preface

Development of wearable devices for human health monitoring has expanded significantly in recent years. In the near future, wearable technologies are expected to become an inseparable part of daily life, enhancing both health and comfort. Advances in this field now enable real-time health monitoring by physicians and healthcare providers. This capability is particularly important for individuals requiring continuous care, including senior citizens, patients with chronic conditions, pregnant women near delivery, and newborn infants. Ensuring the reliability and comfort of wearable systems requires moving beyond conventional rechargeable batteries toward alternative power solutions. Ideally, wearable devices should harvest energy directly from the human body and operate in a self-powered manner. Converting otherwise wasted body energy into useful electrical power offers a promising path toward achieving this goal.

Among various body energy harvesting approaches, including mechanical, biochemical, and thermal methods, thermal energy harvesting based on thermoelectric conversion has emerged as one of the most promising technologies. The human body continuously dissipates heat to the surrounding environment, creating a persistent temperature difference between the skin and ambient air. Thermoelectric materials and devices can directly convert this temperature gradient into electrical power. These devices are solid-state, scalable, reliable, and contain no moving parts, features that are highly required for wearable applications. The field of wearable thermoelectrics lies at the intersection of materials science, physics, chemistry, mechanical engineering, electrical engineering, thermal sciences, textiles, and biomedical engineering.

This book provides a foundational understanding of thermoelectric materials and devices, focusing on the design considerations, engineering constraints, and performance requirements unique to body heat harvesting. While thermoelectric phenomena have been covered extensively in the literature, a dedicated resource that bridges materials science, device architecture, and thermal management under realistic, on-body operating conditions remains limited. Wearable systems demand different material properties and device designs compared to conventional, commercially available thermoelectric modules. To address this gap, this book systematically examines thermoelectric strategies within the unique constraints of wearable applications, paying particular attention to material selection, device design, flexible and rigid platforms, and thermal interface optimization. In addition, the book highlights current applications of wearable thermoelectric power generators and the design considerations for wearable cooling systems.

My personal motivation for writing this book stems from years of research and discussions with students and colleagues working at the intersection of thermoelectrics and wearable electronics. Although thermoelectric devices show significant potential for body heat harvesting, practical design constraints often limit their real-

 | https://doi.org/10.1515/9781501514814-203

world implementation. This book aims to clarify these challenges and discuss practical strategies to overcome them, outlining realistic pathways toward reliable self-powered wearable electronics. The concepts presented here may also be applied to other thermoelectric systems in contact with the human body, such as localized cooling or warming devices.

This book is intended for students, researchers, scientists, engineers, and industry professionals working in energy harvesting, wearable electronics, advanced materials, and device engineering. It may also serve as a reference for graduate-level courses in thermoelectrics, functional materials, and self-powered systems.

It is my hope that this book will contribute to advancing both fundamental understanding and technological development toward self-powered wearable systems. The goal is to present essential concepts in a clear and accessible manner, making them useful to readers from both academia and industry. A deeper understanding of material behavior, device design, and system integration may support further progress in thermoelectric materials and technologies and strengthen our understanding of structure–property–performance relationships in self-powered devices.

I am grateful to my students, colleagues, collaborators, and mentors whose research efforts and discussions have shaped many of the ideas presented in this book.

1 Materials and devices for energy harvesting technologies from human body

Various smart wearable electronic devices, featuring miniaturized sizes and diverse functionality, have recently been developed to monitor physical activity and health using a range of integrated sensors. Powering these sensors and health monitoring aids has been a persistent challenge since their initiation, primarily because batteries are finite energy resources that require frequent charging, which interferes with continuous data acquisition. Moreover, the inherent trade-off between battery size and lifespan underscores the need for identifying a reliable alternative energy source for wearable devices.

Human body energy, including mechanical, biochemical, and thermal energy, offers an unlimited resource for wearables and is a potential inexhaustible replacement for batteries, providing sustainable and clean power for both implantable and wearable devices. The amount of power generated by these body energy harvesters depends on their size, efficiency, and their position on the body. Other factors such as flexibility, weight, shape, and comfort also govern their implementation. This chapter will therefore discuss and compare various energy harvesting techniques derived from the human body.

1.1 Wearable devices

Wearable devices are electronic systems or components designed to be directly or indirectly in contact with the human body (or an animal's body). This contact can be continuous or noncontinuous. These devices are worn on or implanted in the body, and can include items integrated with the skin, embedded in clothing, or functioning as accessories such as backpacks, glasses, garments, patches, bands (e.g., wristbands and armbands), rings, and even personalized chairs and beds. Wearable devices possess a wide range of capabilities, including sensing, monitoring, tracking, data collection, control, communication, responsiveness, energy harvesting, healing, and providing body comfort. Their applications span various fields, including healthcare (especially monitoring), robotics, entertainment, fitness, the Internet of Things (IoT), safety, agriculture, and human–machine interaction.

Holistic wearable healthcare devices are designed to continuously collect, monitor, analyze, save, and transfer essential health data from humans or animals. This data includes measurements such as blood pressure, glucose level, heart rate, electrocardiography (ECG), electroencephalography (EEG), electromyogram (EMG), photoplethysmography (PPG), hydration/sweat, blood oxygen saturation, temperature, and respiration rate in human or animals. Such data can be directly transferred to a medical doctor for further health monitoring (Figure 1.1). Therefore, continuous data collec-

 | https://doi.org/10.1515/9781501514814-001

tion from the body is essential to ensure accurate and uninterrupted health monitoring.

Devices that require frequent battery charging or operate only with continuous plug-in cords have functional limitations that impede their widespread application. In order to overcome this challenge, three main solutions are feasible: (1) The development of low-power sensors that require less energy, enabling longer-lasting battery charges; (2) the fabrication of new solid-state batteries that provide extended charges, eliminating frequent charging and data collection discontinuities; and (3) the implementation of standalone devices powered by body energy. While several industries are actively pursuing the first two solutions, the third solution presents a more practical approach to addressing this challenge. The development of battery-free or less battery-dependent wearable devices has gained significant interdisciplinary research focus since the early 2010s. Future health monitoring devices will require continuous energy harvesting technologies that can either replace the current rechargeable batteries entirely or continuously use body energy to charge them over time [1].

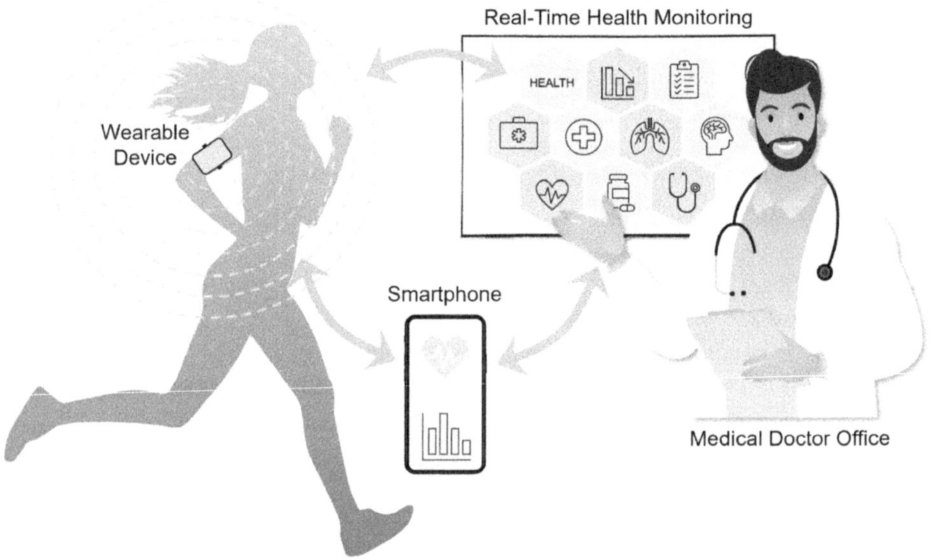

Figure 1.1: A wearable device continuously collects, analyzes, saves, and transmits physiological data, enabling real-time or remote monitoring by a medical doctor.

1.2 Body energy harvesting techniques

The human body provides several forms of energy that can be harvested to power wearable devices. These sources include, but are not limited to, thermal energy (body heat), mechanical energy (motion and pressure), ambient light (through the skin), electrochemical energy (from body fluids), and electrical activity (from biopotential waves and signals). Cumulative energy from daily activities such as walking, jogging, running, swimming, and metabolism is enough to power several low-power sensors and electronic devices. The availability and viability of these body energy resources depend on various factors, such as age, activity level, room temperature, and ambient light.

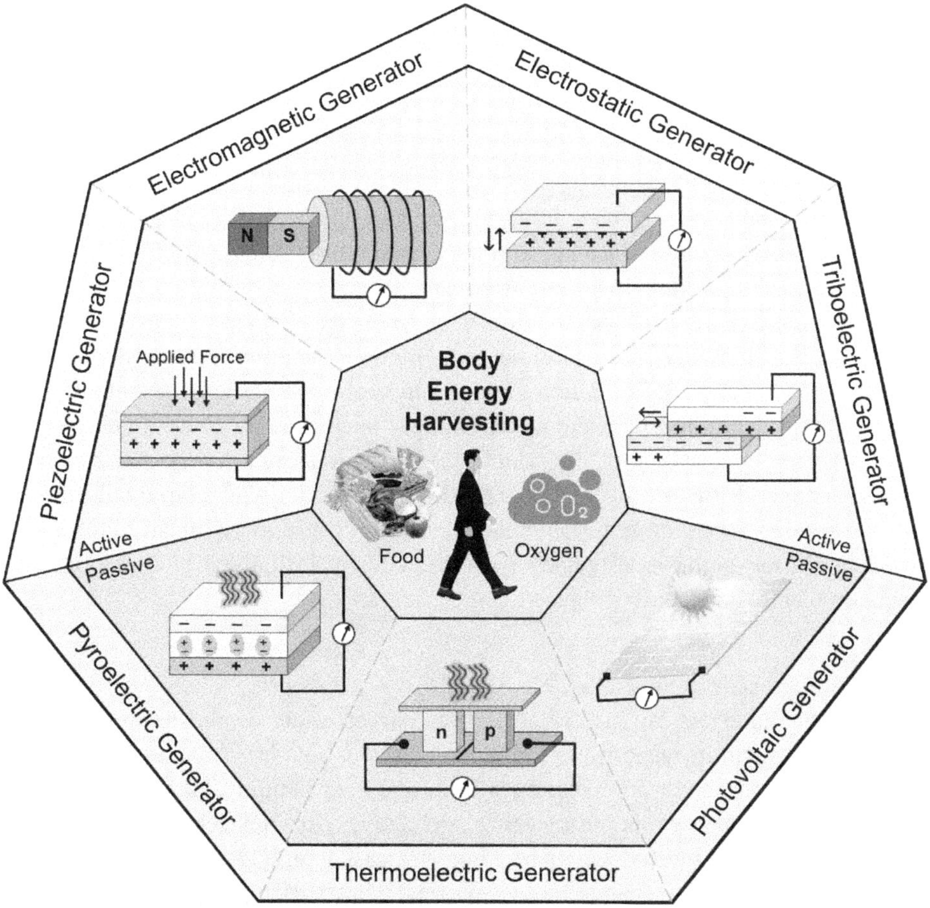

Figure 1.2: Active and passive energy harvesting from the human body can both be utilized for wearable applications. Active energy harvesting methods leverage kinetic energy and include piezoelectric, electromagnetic, electrocaloric, and triboelectric generators. Passive energy harvesting methods, on the other hand, utilize heat or ambient light, including pyroelectric, photovoltaic, and thermoelectric generators. The basic concepts and working principles of these energy harvesters are illustrated.

Body energy harvesting techniques are generally divided into active and passive mechanisms (Figure 1.2). An active energy harvester generates energy through intentional operation or by enhancing the body's physical activity and kinetic energy, such as through mechanical interaction or movement. Examples of active energy harvesters include, but are not limited to, piezoelectric, electromagnetic, electrostatic, and triboelectric generators. Conversely, a passive energy harvester produces energy from the body's natural physiological processes without the use of secondary stimuli. Examples of passive energy harvesters include, but are not limited to, pyroelectric, photovoltaic, and thermoelectric generators. According to this definition, active and passive energy harvesters can be interrelated. For instance, users can utilize forced airflow to enhance the performance of thermoelectric (TE) systems.

1.2.1 Active energy harvesters

Active energy harvesters convert low-level ambient kinetic energy into useful forms, such as electricity [2]. The efficiency and maximum power generated by these devices depend on several factors, including materials properties, device design, the energy conversion system, response time and frequency, the specific energy harvesting technique, and the characteristics of human body motion, including its time domain and impact on body comfort. The human body's physical activities such as walking, jogging, running, swimming, organ motions, eating, chewing, and talking can all serve as sources for energy harvesting. The viability of active energy harvesters in wearable platforms depends on several key factors: their intended application, flexibility, the type of applied force, system robustness, weight, cost, and required voltage and power levels. Key active energy harvesting mechanisms include piezoelectric, electromagnetic, electrostatic, and triboelectric approaches.

1.2.1.1 Piezoelectric generators

Piezoelectric materials can generate electric charges under a mechanical load, and conversely, deform in response to an applied electric field [3]. This mechanical load can be in the form of vibration, compression, stretch, or bending. Piezoelectric generators (PZGs), also known as piezoelectric transducers, are devices that harness this property, converting mechanical energy into electrical energy. Alternatively, they can utilize an applied electric field to induce material deformation [4, 5].

PZGs stand as the most reliable mechanically induced energy harvesting technique, offering the highest generated power density compared to other active energy harvesting methods. These devices effectively extract and convert ambient human body motions into electricity. PZGs have high energy density, exhibit small mechanical damping, possess an inherent reciprocal conversion capability, and facilitate the fabrication of scalable architectural micro-electromechanical systems (MEMS) [4, 6].

In wearable platforms, depending on the direction of electric field and mechanical strain, PZGs can mainly work in two modes: d_{31} (or transverse mode) and d_{33} (or longitudinal mode). In d_{31} mode, mechanical stress is applied perpendicular to the direction of the electric field produced. Piezoelectric energy harvesters most commonly utilize cantilever beam structures in the d_{31} mode. In d_{33} mode, the mechanical stress and the generated electric field act in the same direction, typically along the thickness of the material. Stacked transducers are characteristic examples of harvesters utilizing the d_{33} mode, making them well-suited for pressure-mode operation. Several design constraints contribute to challenges in PZGs. These include the critical need for frequency matching between the ambient vibration and the harvester's natural frequency, as well as managing their inherently high output impedance. Furthermore, the power generated by PZGs is inversely proportional to the material's damping ratio, which must be minimized through careful materials design and fabrication. Optimal PZGs also require low dielectric permittivity, high mechanical durability, and high piezoelectric coefficients [6].

While PZGs enable compact and flexible energy harvesters with high energy density when sufficient motion is provided, they do face limitations. These include material fatigue or fracture over time, temperature sensitivity, diminished power output at low strain or stress levels, and the need for AC power rectification. Nonetheless, they are well-suited for applications such as joint movement and footstep harvesting, as well as integration into smart textiles and pads. PZGs for wearable applications are mainly fabricated from materials like lead zirconate titanate (PZT), barium titanate ($BaTiO_3$), polyvinylidene fluoride (PVDF), zinc oxide (ZnO), potassium sodium niobate (KNN), and lead magnesium niobate–lead titanate (PMN-PT). The power output generated from wearable PZGs typically ranges from a few microwatts to a few hundred microwatts, depending on the specific material, mechanical force, device geometry, and application location [7, 8].

1.2.1.2 Electromagnetic or magnetic induction generators

Magnetic induction generators (MIGs), which operate according to Faraday's law of electromagnetic induction, represent another practical approach for harvesting energy from ambient vibrations of the human body. In this method, mechanical movement induces an electromotive force (EMF) through the motion of a permanent magnet (such as such as NdFeB, CoPt, FePt, or SmCo) relative to a copper coil [9]. While MIGs offer a long lifespan and can generate power from very low resonance frequencies [10], they demonstrate low power density and output voltage, which can be mitigated by increasing the magnetic flux density across the coil [9]. When integrated into wearable platforms, MIGs present challenges, including low efficiency, miniaturization limitations, ohmic losses, and the bulkiness of the required magnets.

MIGs can be effectively deployed in devices like wristbands, armbands, knee-joint braces, or footwear to generate power from mechanical movement. Depending on the specific device, form factor, activity level, motion type, and design, MIG devices can produce power ranging from a few microwatts (μW) to several milliwatts (mW) [11]. However, the power density of these devices, typically in the range of 0.001–10 mW/cm^3, is often marginal for consistently powering low-power electronic sensors [12–17].

1.2.1.3 Electrostatic generators

Electrostatic generators (ESGs) offer another effective method for converting mechanical energy into electricity. An ESG is essentially a capacitive device composed of two separated, conductive parallel plates (e.g., aluminum, gold, or graphene) where electrical charge is stored. The space between these plates is filled with either air or a dielectric material (e.g., Teflon, PDMS, or PET). Energy is generated by moving one plate relative to the other via an external mechanical stimulus, which alters the plate separation and/or effective area, thereby inducing a change in capacitance. This variation alters the electric potential and stored electrostatic energy, converting the mechanical work done on the capacitor into harvestable electrical energy [18–21].

The primary advantages of ESGs include their capability for high voltage generation, reduced dependency on specific mechanical resonant structures, and broadband operational flexibility. However, they mainly suffer from issues such as dielectric breakdown, high output impedance, low output current, the need for an external voltage source for startup [22, 23]. Depending on their shape, size, and type, ESG devices can generate power ranging from nanowatts (nW) to hundreds of microwatts (μW) [19, 24, 25]. Accordingly, current ESG technologies are generally insufficient to power standalone wearable nodes independently, often requiring integration with complementary mechanical energy harvesting techniques.

1.2.1.4 Triboelectric generators

Triboelectric generators (TGs), also known as triboelectric nanogenerators (TENGs), represent the newest mechanical energy harvesting techniques for ambient human body motions, operating based on triboelectric effect. This effect refers to the generation and accumulation of static electricity when the surface of two distinct (dielectric) materials are rubbed together and subsequently separated. One of these surfaces acts as an electron donor, while the other behaves as an electron acceptor. Consequently, the application of mechanical stimuli generates equal and opposite charges that appear on the surfaces of the two materials [26–28]. The positive surfaces are usually made of materials such as nylon, mica, silk, wool, fur, cellulose, and skin, while negative surfaces are typically made of PTFE, FEP, PDMS, PP, and PVC.

The polarity and density of these charged particles depend on various factors, including material properties, device design, applied stress, surface morphology and roughness, temperature, and environmental conditions [27]. TGs feature diverse surface morphology and features, ranging from millimeter to nanometer patterns such as lines, hemispheres, pyramids, cubes, and squares. The advantages of TGs include high voltage output, low-frequency requirements, lightweight design, flexibility, scalability, low cost, and simple fabrication techniques. However, their primary limitations include low efficiency, low current output, sensitivity to humidity, and surface charge saturation, in issues related to mechanical fatigue and wear.

These unique properties offer extensive applications for TGs in wearable devices, especially considering that human skin itself can act as a triboelectric material [29]. The combination of high voltage (ranging from a few volts to hundreds of volts) and low current (in the microampere range) necessitates power-conditioning stages, including rectification and energy storage elements like supercapacitors, to ensure efficient power delivery to electronics. In wearable platforms operating at low frequencies, the generated power typically ranges from microwatts (μW) to milliwatts (mW), with voltages spanning over 100–500 V [30–32].

1.2.2 Passive energy harvesters

Passive energy harvesters are devices designed to generate energy from natural and instinctive body processes or the environment without requiring external stimuli or additional effort. The primary energy resources for these systems are ambient environment and heat. The viability of passive energy harvesters in wearable platforms depends on several factors, including their intended applications, flexibility, environmental conditions, system robustness, weight, costs, and the required voltage and power levels. Key passive energy harvesting mechanisms include pyroelectric, photovoltaic, and thermoelectric approaches.

1.2.2.1 Pyroelectric generators

Pyroelectric generators (PEGs) utilize temperature fluctuations over time ($\Delta T/\Delta t$) to produce electricity by altering the orientation of dipoles within the pyroelectric material. PEGs generate voltage when they are heated or cooled because changes in temperature adjust the positions of these dipoles, inducing internal displacements within the crystal lattice and consequently altering its polarization. The resulting voltage persists as long as the temperature continues to fluctuate. The harvested energy can then either be used directly to power electronics or stored in capacitors. [33].

Notable pyroelectric materials include ferroelectric ceramics such as PZT, $BaTiO_3$, KNN, and lithium tantalate ($LiTaO_3$), as well as polymers such as PVDF, and composites such as PVDF + $BaTiO_3$. The pyroelectric coefficient and Curie temperature are two main contributing factors in material selection for PEGs [34]. The Curie temperature is defined as the temperature above which a pyroelectric material loses its spontaneous polarization and can no longer generate electricity under an applied electric field.

PEGs are capable of harvesting energy from low-frequency heat changes, making them ideal for slow cyclic temperature variations, such as those similar to human skin condition. However, they critically require transitional temperature fluctuations rather than steady-state conditions. PEGs typically generate low power, usually in the range of a few microwatts [35–38].

1.2.2.2 Photovoltaic generators

Photovoltaic generators (PVGs), commonly known as solar cells (SCs), are solid-state devices that convert ambient light (photons) directly into electricity through the photovoltaic effect. This process occurs when photons interact with a semiconductor material, transferring energy to its electrons. This energy absorption excites electrons, creating electron-hole pairs and, consequently, a flow of current [39].

PVGs are solid-state and lightweight devices with no moving parts, offering a long operational lifetime and requiring low maintenance. They can continuously generate power whenever light, whether from sunlight or artificial lamps, is available. For wearable applications, particularly those demanding surface areas greater than a few square centimeters, flexible PVGs are an ideal choice. These are typically fabricated from organic, perovskite, or thin-film SCs, allowing for integration into clothing or direct application on the skin. The application of PVGs in wearables faces significant challenges. Their output power is highly dependent on light exposure and orientation, leading to significant drops in performance in shaded conditions, low ambient light, or at night. Consequently, they cannot serve as an uninterrupted power source on their own, thus requiring an integrated battery or supercapacitor to ensure a steady power supply. Furthermore, current flexible PVGs are prone to instability due to their sensitivity to moisture and heat [40, 41].

While PVGs generally offer higher conversion efficiencies compared to other energy harvesting technologies, this advantage is often weakened in wearable applications. For instance, despite impressive efficiencies of over 47% in multijunction cells [42] and over 19% in organic cells [43], their wearable performance is restricted by the need for sufficient ambient light and their reduced efficiency under typical indoor lighting conditions [44].

Wearable applications for PVGs are diverse, including, but not limited to, health monitoring devices, skin patches, smartwatches, fitness bands, smart textiles and flexible fabrics, eyewear, outdoor wearables, and backpacks. The required surface area for these applications varies significantly, ranging from 2 to 5 cm^2 for smartwatches and skin patches to approximately 100–400 cm^2 for backpacks. PVG devices can generate power ranging from a few to hundreds of microwatts (μW) in indoor environments. Research into PVGs for wearable and implantable devices remains an active field. Current efforts primarily focus on optimizing device structures and materials, as well as achieving high conversion efficiencies under indoor lighting conditions [44–49].

1.2.2.3 Thermoelectric generators

Thermoelectric (TE) materials, typically semiconductors, can generate electricity when a temperature difference is applied across them, and vice versa. TE generators (TEGs) are solid-state devices composed of TE materials that can directly convert heat into electrical energy [50].

The human body continuously generates heat, producing approximately 100 W while at rest and up to 400 W during physical activity. Given the average body surface area of an adult woman (approximately 16,000 cm^2) and man (approximately 19,000 cm^2), this translates to a heat generation rate of 5.3 to 25 mW/cm^2. Since low-power wearable electronics typically require about 500 µW to 10 mW of power, TEGs can effectively harvest this thermal energy and convert it into electricity.

TEGs are lightweight, maintenance-free, and operate continuously without requiring any mechanical motion. They function efficiently both indoors and at night, even when the user is at rest. These characteristics have established TEGs as one of the most reliable methods for enabling self-powered wearable devices, significantly reducing the need for frequent battery charging. Their uninterrupted operation further makes them appealing for continuous monitoring applications. The most effective wearable TEGs are often fabricated from bismuth telluride (BiTe)-based compounds [51]. In this book, BiTe and Bi_2Te_3 are used interchangeably to refer to bismuth telluride alloys. Depending on the material's performance, as well as the device and system design, TEGs can generate power ranging from a few microwatts to

Table 1.1: Comparison of state-of-the-art human body energy harvesting mechanisms.

Method	Power density	Voltage	Size (area)	References
Piezoelectric	64.9 µW/cm^2	3 V	1.54 cm^2	[52]
Magnetic induction	507 µW/cm^3	3 V	3.76 cm^3	[53]
Electrostatic	1.2 µW/cm^2	40 V	1 cm^2	[54]
Triboelectric	50 mW/cm^2	500 V	60 cm^2	[55]
Pyroelectric	2.86 µW/cm^2	10 V	52.4 cm^2	[56]
Photovoltaic	2.2 µW/cm^2	0.2 V	–	[57]
Thermoelectric	44.2 µW/cm^2	129 mV	0.9 cm^2	[51]

Table 1.2: Source of power, advantages, and limitations of different energy harvesting methods.

Method	Source of power	Advantages	Limitations
Piezoelectric	Body motion/ strain	– Smaller sizes and thin films – High power density and output voltage – Flexibility – Simple construction – High sensitivity – Compatibility with MEMS fabrication	– Requires mechanical motion – Output is AC – needs rectification – Brittle materials – Low efficiency at low frequency – High impedance
Electromagnetic (or magnetic induction)	Body motion/ swing	– High output current and power – Low impedance – Simple structure – Good for joint or limb movement – Long lifespan	– Low efficiency – Low output voltage – Sensitive to vibrational frequency – Less flexibility and scalability – Difficulties in miniaturization – Large ohmic losses – Need for bulky magnets and coils
Electrostatic	Motion + variable gap	– Compatible fabrication process – High power density – High output voltage – Suitable for low-frequency operation – Lightweight	– Need for separate voltage source for initial charge-up – High voltage insulation – Parasitic components – Leakage current issues – High output impedance – Low output current – Sensitive to packaging and environment

Triboelectric	Friction/ contact	– Very high power density – High output voltage – Low-frequency requirements – Lightweight – Excellent flexibility – Simple fabrication – Scalability – Low cost – Maintenance-free	– Low efficiency – Low current output – Lack of fundamental understanding – Dependent on material and surface properties – Sensitive to humidity – Surface charge saturation – Mechanical fatigue and wear
Pyroelectric	Temperature fluctuation	– Ideal for slow cyclic temperature variations such as body skin – Scalability – No heat sink	– Low output power density – Low efficiency – Less pyroelectric materials – Complex operation – Limited frequency of operation – Require transitional temperature fluctuations – No power with constant temperature
Photovoltaic	Light (sun or indoor light)	– Low cost – Lightweight – No moving part – High energy under light – Easy fabrication and integration into wearables – Maintenance-free – High reliability – Long lifetime – Scalability – Flexibility	– Limited indoor efficiency – Location dependent device performance – Dependent on light (no power in dark) – Needs surface exposure
Thermoelectric	Body heat	– Continuous operation – High power density	– Contact resistance – Low efficiency

(continued)

Table 1.2 (continued)

Method	Source of power	Advantages	Limitations
		– Low cost – Lightweight – No moving part – Maintenance-free – High reliability – Longer lifetime – Scalability – Simple construction – Flexibility	– Low voltage – May need voltage boost converter – Performance reduction in flexible devices – Low ΔT on skin limits performance – Needs heat sink

several hundreds of microwatts. A more detailed explanation of the concepts, physics, and applications of TEGs will be provided in the next chapter.

Tables 1.1 and 1.2 represent the comparison of the state-of-the-art human body energy harvesters, along with their advantages and challenges. In the following chapters, we will evaluate the principles of thermoelectricity and wearable TEGs and their applications in addition to address the challenges and potential improving strategies.

2 Principles and applications of thermoelectric materials

The thermoelectric (TE) effect refers to the direct conversion of a temperature gradient across a material into voltage (or current). This phenomenon is reversible: an applied electrical current can generate a temperature gradient across a TE material, resulting in cooling. This unique capability stems from the distinctive electronic and thermal transport properties inherent in these materials. A comprehensive understanding of TE phenomena and material requirements is essential to advancing this technology for various applications. This chapter discusses the principles of thermoelectricity and its general applications.

2.1 Principles of thermoelectricity

When a temperature difference exists across a TE device, a voltage is generated, and vice versa. At the atomic scale, the underlying mechanisms depend on the operating mode: under a temperature gradient, charge carriers (electrons or holes) diffuse from the hot side to the cold side. Conversely, when an external electric field is applied, these carriers drift along or against the direction of the field depending on their charge, driving localized heating or cooling. The TE phenomenon is introduced by three fundamental concepts known as the Seebeck (α or S), Peltier (Π), and Thomson (τ) effects.

Suppose that conductors A and B are connected at both ends, named as junctions C and D (Figure 2.1). Conductor A is divided into two free parts for further characterization. A temperature difference ($\Delta T = T_2 - T_1$) is applied between the junctions while the free parts of A are maintained at the same temperature. A voltage difference (ΔV) appears between free ends. The thermopower or the Seebeck coefficient (α_{AB} or S_{AB}) is defined as follows:

$$\alpha_{AB} = \frac{\Delta V}{\Delta T} \tag{2.1}$$

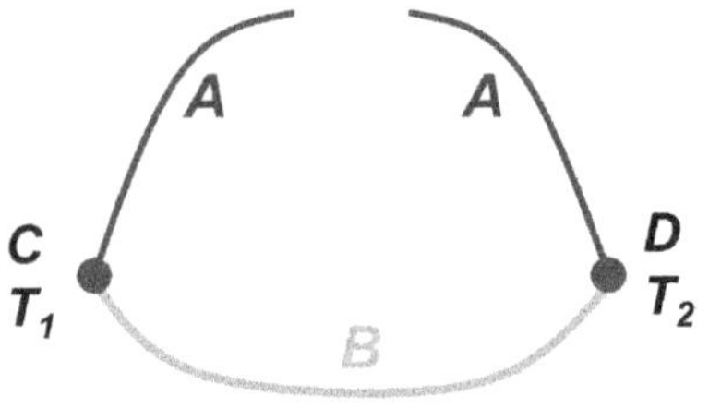

Figure 2.1: Basic of a thermocouple: A and B are conductors.

 | https://doi.org/10.1515/9781501514814-002

On the contrary, the differential Peltier coefficient (Π_{AB}) is defined when an electric current (I) is applied at free ends. In a clockwise current direction, the positive Peltier coefficient results in the junction C being heated and D cooled. Π_{AB} (eq. 2.2) is defined as the ratio of heat exchange (heating or cooling) at the junctions C and D to the electric current (I):

$$\Pi_{AB} = \frac{Q}{I} \tag{2.2}$$

According to the Kelvin relations, the thermopower, the TE cooling (Π_{AB}), and temperature (T) are related according to the following equation:

$$\Pi_{AB} = \alpha_{AB} T \tag{2.3}$$

The Thomson coefficient is defined as the rate of heat generation per unit volume of a conductor when an electric current flows through it under a unit temperature gradient. According to the Kelvin relation, the Thomson coefficient is defined as follows:

$$\tau_A - \tau_B = T\frac{d\alpha_{AB}}{dT} \tag{2.4}$$

It is worth noting that α is a physical property and can be measured while Π and τ are calculated accordingly.

2.2 Properties of thermoelectric materials

Considering three fundamental effects (Seebeck, Peltier, and Thomson), there is a group of counteracting properties, including electrical conductivity (σ) , Seebeck coefficient (α or S), and thermal conductivity (κ). These properties need to be optimized for a high-performance TE material, which will be discussed here in details [58].

2.2.1 Thermoelectric figure of merit, *zT*

The dimensionless figure of merit, zT, determines the TE performance of a material. It is defined by material's intrinsic transport properties, namely electrical conductivity (σ), thermopower or the Seebeck coefficient (α) and thermal conductivity (κ) at a given absolute temperature (T):

$$zT = \frac{\sigma\alpha^2}{\kappa} T \tag{2.5}$$

The term $\sigma\alpha^2$ is called the power factor. To maximize zT of a material, one has to increase the electrical conductivity and the Seebeck coefficient, while decreasing the thermal conductivity; in other words, zT is the measure of the competition between electronic transport and thermal transport in a material. Since each feature has different interrelated properties, many parameters must be optimized to maximize zT [58, 59].

2.2.2 Electrical conductivity

Electrical conductivity (σ) or specific conductance is the ability of a material to carry the electrical current. It has physical relationships with electrical resistivity (ρ), carrier concentration (n), elementary charge (e), and mobility of carriers (μ), according to the following equation:

$$\sigma = 1/\rho = ne\mu \tag{2.6}$$

In metals, electrical conductivity is very high due to the presence of many carriers and states (~10^{22} carriers per cm^3) for conduction. In semiconductors, electrical conductivity is dominantly dependent on carrier concentration and mobility, both of which are functions of temperature. This electrical conductivity can be increased either through chemical doping or by optimizing carrier mobility [60]. In insulators, electrical conductivity is negligible due to the very high bandgap that limits the number of carriers.

2.2.3 Seebeck coefficient

The Seebeck effect is an intrinsic property of a material where the voltage gradient is induced by the temperature gradient [61]. As mentioned earlier, the thermopower or the Seebeck coefficient (α or S) is defined as the ratio of the generated voltage to the temperature difference ($\alpha = \Delta V/\Delta T$). The Seebeck coefficient is very low for metals (a few μV/K) and much higher for semiconductor materials (~ a few hundred μV/K). In many TE studies, the Seebeck coefficient is interpreted as a measure of transport entropy per carrier, which is proportional to the average energy of the carriers relative to the Fermi energy (E_F) divided by ($\pm e$) and temperature (T) [50, 62]. From a simple model of electron transport, the Seebeck coefficient has the following interrelationships with carrier concentration (n) and effective mass (m^*) for metals and degenerate semiconductors [58]:

$$\alpha = \frac{8\pi^2 k_B^2}{3eh^2} m^* T \left(\frac{\pi}{3n}\right)^{2/3} \tag{2.7}$$

Good TE materials should have a high Seebeck coefficient. Due to the direct relationship between α and m^*, a higher density of states effective mass increases the Seebeck coefficient, but it may also lead to a high conductivity effective mass and reduce the carrier mobility; hence, the electrical conductivity. Therefore, optimizing the effective mass tensor is essential to achieve a balance between a high Seebeck coefficient and sufficient carrier mobility.

2.2.4 Thermal conductivity

Thermal conductivity (κ) is the ability of a material to conduct heat. The primary mechanism for heat conduction in semiconductors and insulators is the vibration of the crystal lattice, quantized as phonons. Consequently, heat flow is governed by the propagation and scattering of phonons across different wavelengths and mean free paths. Thermal conductivity originates from three distinct transport mechanisms: first, the electronic component, representing electrons and holes transport heat (κ_e), second, the lattice component, representing phonons that carry heat through the lattice (κ_l), and third, mixed carrier conduction, known as the bipolar component (κ_{bi}). According to theWiedemann–Franz law, there is a direct relationship between the electronic term and electrical conductivity; hence, the carrier concentration, and mobility:

$$\kappa_e = L\sigma T = ne\mu LT \tag{2.8}$$

where $L = 2.4 \times 10^{-8}$ J^2/K^2 C^2 is the Lorentz factor for free electrons. One may calculate κ_e from the electrical conductivity data and estimate the κ_l by $\kappa - \kappa_e$, provided that κ_{bi} is negligible. However, in the cases of low carrier concentration, L may reduce by as much as 20% of the free-electron value. Besides, when there is κ_{bi} due to the mix conduction, κ_{bi} can be comparable with κ_e and κ_l. Therefore, there is a source of mistrust in κ_e estimations using the Wiedemann–Franz law [63].

To optimize TE efficiency and maximize zT, high electrical conductivity and low thermal conductivity are often required. However, as illustrated by eq. (2.5), meeting both conditions simultaneously presents a significant challenge. One practical approach to improving zT in semiconductors is energy band structure engineering, as both the Seebeck coefficient and electrical conductivity are functions of the band characteristics and doping level. Furthermore, thermal conductivity can be reduced through chemical substitutions and by minimizing lattice thermal conductivity. To minimize lattice thermal conductivity, phonon transport must be impaired without deteriorating electrical conductivity. Nanostructuring techniques have helped scientists to overcome this challenge by scattering phonons at different frequencies through various methods. These methods include rattling scattering, grain boundary scattering due to the interfaces, mass fluctuation scattering and interface scattering in alloys, heterostructures, nanocomposites, thin films, and multilayer systems [60, 64, 65].

A single TE material is rarely used in its synthesized form. In practice, the performance of TE materials is evaluated when they are incorporated into a device and integrated into an electronic system. The components, fabrication process, and characterization of TE devices will be discussed in Chapter 3.

2.3 Applications of thermoelectrics

Since the discovery of the TE effect by Thomas Johann Seebeck in 1821, the applications of TE materials and devices have expanded extensively. The Russians were early pioneers in their practical use, being the first to employ TE generators (TEGs) to power radio stations in the 1920s. Kerosene radios became popular after World War II. In rural areas lacking electricity, burning kerosene not only provided light but also generated up to 90 V and 2 W of power to run an all-wave portable radio (Figure 2.2a) [66]. Since then, the use of thermoelectrics has continued to expand into various fields due to their unique characteristics, including a lack of moving parts, quiet operation, compactness, reliability, environmental friendliness, low maintenance, and dual functionality.

The capability of TE devices is vast; however, their performance is fundamentally constrained by material properties and specific applications. Future innovations in materials science and device design will be critical to extending and stabilizing their use across various industries. In recent years, several companies have focused on innovative products based on TE concepts. While some of these ventures have been successful, others have been discontinued. What is clear is that biomedical, wearable, and body-contact surface applications of thermoelectrics have been growing extensively, underscoring the increasing recognition of the technology's importance.

2.3.1 Aerospace applications

Since 1961, TEGs have played a crucial role in powering long-duration space missions. NASA began launching spacecraft equipped with radioisotope thermoelectric generators (RTGs). The US Navy's Transit 4A and 4B satellites were the first to utilize this technology. In the 1970s, RTGs enabled deep space explorations. Pioneer 10 (1972) and Pioneer 11 (1973) explored Jupiter and Saturn, and Voyager 1 and 2 (1977) remain operational in interstellar space today (Figure 2.2b). Other spacecraft powered by RTGs include Galileo (1989), Ulysses (1990), Cassini (1997), New Horizons (2006), the Curiosity rover (2011), and the Perseverance rover (2020).

2.3.2 Automotive industry

Since TEGs can convert waste heat directly into electricity, they have found promising applications in the automotive industry. Primarily, this involves generating power from exhaust and engine heat, thereby improving overall fuel consumption efficiency. For instance, BMW, Honda (Figure 2.2c), and Ford have used TEGs for this purpose in their hybrid systems, saving up to 5% fuel. Another application of TE is in localized or cabin climate control, where TE devices replace conventional HVAC components. Specifically, TE coolers and heaters are integrated into car seats and steering wheels for fast, localized temperature regulation. Additionally, in electric and hybrid vehicles, TE enables precise battery thermal management to enhance both performance and longevity.

2.3.3 Solid-state coolers

The electrical to thermal conversion mode of the TE effect, known as the Peltier effect, was discovered by Jean Peltier in 1834. Thermoelectric coolers (TECs) were invented in the mid-1950s following the discovery of bismuth telluride (Bi_2Te_3), the most prominent TE material to date. The application of TECs began with mini-fridges and beverage coolers. Due to their unique characteristics, such as silent operation, compactness, and precise temperature control, their current applications have expanded to various fields such as electronics, medical devices, and aerospace. These applications include, but are not limited to, electronic cooling (such as CCD cameras, CPUs, infrared detectors, laser diodes, optical sensors, satellite instruments, and sensors), battery thermal management, portable vaccine and organ transport, and wearable coolers.

2.3.4 Emerging applications

In recent years, the applications of TE devices have expanded enormously, with many novel uses being proposed. This significant growth is largely due to ongoing advancements in materials, processing, and fabrication techniques, which enable these devices to provide sustainable energy resources and effective thermal management solutions. The emerging applications for TEs are diverse, ranging from waste heat recovery in power plants, nuclear reactors, and low-grade heat sources, to cutting-edge fields such as robotics, pharmaceuticals, data center cooling, smart textiles, and wearables. TEs are also finding roles in solar-thermal generators, smart buildings, Internet of Things (IoT) devices, HVAC systems, and self-powered sensors.

Figure 2.2: Applications of thermoelectric generators: (a) kerosene radio and lamp generator (reproduced from Ref. [66], with permission from Elsevier), (b) Voyager spacecraft before its launch (image courtesy of NASA [67]), and (c) initiated by BMW [68], TEG system for waste heat recovery in automotive industry in Honda prototype. Reproduced from Ref. [69], with permission from Elsevier.

3 Thermoelectric materials and devices for wearable applications

To date, thermoelectric (TE) materials and devices represent a leading technology for harvesting energy from the human body. Over the last three decades, extensive research has been conducted across academia and national laboratories to enhance the performance of TE materials, while industry has focused primarily on device scaling and manufacturing. Historically, both sectors targeted conventional, rigid TE systems. More recently, driven by advancements in low-power microelectronics and smart health sensors, the integration of TE materials into wearable devices has gained significant momentum. However, the material properties and architectural designs required for body-heat harvesting differ fundamentally from those used in traditional power generation applications. This chapter provides a foundational overview of TE materials, device fabrication methodologies, and standard testing methods.

3.1 Thermoelectric materials

TE materials have been actively discovered and optimized over the last few decades. Since 2010, the number of discoveries has increased exponentially, and several high-performance and promising TE materials have been found or improved. However, a primary challenge facing many of these recent high-zT materials is their instability and discrepancies regarding characterization accuracy, both of which require rigorous further investigation.

There are various types and categories of TE materials, each with different transport properties and working temperature ranges. Some of these materials have been widely developed and are commercially used, while others have been developed only on a laboratory scale. These materials include, but are not limited to:

i) Chalcogenides and their compounds: There are chalcogenides of various elements like bismuth, antimony, lead, tin, copper, germanium, and silver. Examples include bismuth telluride (Bi_2Te_3), bismuth selenide (Bi_2Se_3), bismuth sulfide (Bi_2S_3), antimony telluride (Sb_2Te_3), antimony selenide (Sb_2Se_3), antimony sulfide (Sb_2S_3), lead telluride (PbTe), lead selenide (PbSe), lead sulfide (PbS), tin telluride (SnTe), tin selenide (SnSe), tin sulfide (SnS), copper selenide ($Cu_{2-x}Se$), copper sulfide (Cu_2S), copper telluride ($Cu_{2-x}Te$), germanium telluride (GeTe), germanium selenide (GeSe), germanium sulfide (GeS), silver selenide (Ag_2Se), silver sulfide (Ag_2S), and silver telluride (Ag_2Te).

ii) Antimonide compounds such as bismuth antimonide ($Bi_{1-x}Sb_x$) and lead antimonide ($PbSb_2$).

 | https://doi.org/10.1515/9781501514814-003

iii) Silicide compounds such as silicon germanium (SiGe), magnesium silicide (Mg_2Si), iron silicide ($FeSi_2$), chromium silicide ($CrSi_2$), and high manganese silicide ($MnSi_x$).
iv) Zintl phases such as ytterbium manganese antimonide ($Yb_{14}MnSb_{11}$), calcium zinc antimonide ($Ca_9Zn_{4+x}Sb_9$), and magnesium bismuthide (Mg_3Bi_2).
v) Skutterudite compounds such as cobalt antimonide ($CoSb_3$), nickel antimonide ($NiSb_3$), and iron antimonide ($FeSb_3$).
vi) Half-Heusler (hH) compounds such as titanium-based hHs (TiNiSn, TiCoSb, and TiFeSb), zirconium-based hHs (ZrNiSn, ZrCoSb, and ZrFeSb), and hafnium-based hHs (HfNiSn, HfCoSb, and HfFeSb).
vii) Oxide compounds such as layered cobaltites (like sodium cobalt oxide ($NaCo_2O_4$) and calcium cobalt oxide ($Ca_3Co_4O_9$)), perovskite oxides (like strontium titanate ($SrTiO_3$) and calcium manganate ($CaMnO_3$)), zinc oxide (ZnO), and copper oxide (CuO).
viii) Organic compounds such as conductive polymers (like poly(3,4-ethylenedioxythiophene) (PEDOT), polyaniline (PANI), and polypyrrole (PPy)), halide perovskites (like $FASnI_3$), and organic–inorganic hybrids (like PEDOT and conductive polymers with nanostructured inorganic fillers).
ix) Other compounds and structures such as quantum dots (like PbTe/PbSe and CdTe/CdSe), nanowires (like Bi_2Te_3 nanowires, PbTe nanowires, SiGe nanowires, and Si nanowires), superlattices (like Bi_2Te_3/Sb_2Te_3 sand PbTe/PbSe), clathrates (such as $Ba_8Ga_{16}Ge_{30}$ and $Sr_8Ga_{16}Ge_{30}$), and tetrahedrites (such as $Cu_{12}Sb_4S_{13}$ and $Cu_{10}Te_4S_{13}$).

Among all these materials, bismuth telluride (Bi_2Te_3)-based TE materials, discovered in the mid-1950s, remain the most prominent, reliable, and widely used for applications ranging from −50 °C to 250 °C, including body heat harvesting for wearables.

3.2 Thermoelectric devices

A TE device consists of an electrical insulator for its top and bottom headers (also known as substrates), interconnects, TE *p*- and *n*-type legs, and solder/brazing materials. Figure 3.1 shows the different components and the fabrication process of a TE device. Headers are typically made of ceramics with high thermal conductivity, such as alumina (Al_2O_3) and aluminum nitride (AlN) (Figure 3.1a), or a direct-bonded copper (DBC) substrate. These headers must be electrically insulating but thermally conductive to transfer heat with minimal resistance. They must also have a small thermal expansion coefficient to minimize asymmetric expansion caused by temperature differences across the device. DBCs are composed of a similar ceramic substrate with a copper sheet bonded to one or both sides. Interconnects, primarily composed of copper or gold-coated copper, are patterned onto the ceramic substrate to provide electri-

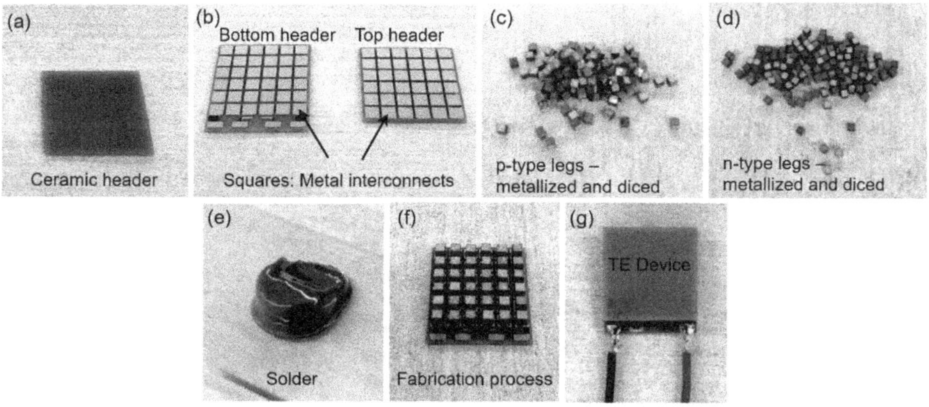

Figure 3.1: Thermoelectric (TE) device compartments and fabrication process: (a) external surface of the header or substrate made of ceramic, (b) top and bottom headers with interconnects, (c and d) *p*- and *n*-type BiTe TE legs metalized and diced, (e) solder paste, (f) fabrication process via pick and place of the legs and subsequent solder bonding, and (g) fabricated TE device [70]. Reproduced with permission from Ref. [70]. Copyright © [2019] American Chemical Society.

cal contact between the TE legs. In the case of DBCs, the copper sheeting can also serve as an interconnect. Interconnects are used to connect individual TE legs in an electrical series (Figure 3.1b).

TE *p*- and *n*-type legs are separately manufactured in bulk form using various techniques such as melting, extrusion, or powder processing. These TE materials can be single crystal, polycrystal, or nanocrystal. The ingots are then cut to a specific thickness, and subsequently metalized (coated) with one or more layers of metals, such as gold or nickel. This metallization minimizes contact resistance between the legs and the interconnects. Finally, the coated materials are diced into specific sizes to produce the final product, known as a TE leg (Figure 3.1c and d).

The legs are connected to the metal interconnects using a solder. In certain applications, such as flexible wearables, the interconnect and solder materials are integrated to form a compliant, reliable metallic junction. Such solders (Figure 3.1e) for wearable modules are commonly made of lead–tin, antimony–tin, gold–tin, or eutectic gallium–indium (EGaIn) alloys. It is important to note that these processes and steps may vary depending on the specific application, required flexibility, and chosen materials. The final fabrication process for the TE device typically involves the pick-and-place positioning of the TE legs, followed by solder reflow bonding (Figure 3.1f). As shown in Figure 3.1g, two wires are soldered to the terminal pads at both ends of the device to serve as the electrical leads for current and power transport.

While the aforementioned steps outline standard TE device fabrication protocols, each stage can be optimized to accommodate specific application demands and material constraints. The TE module features several interfaces between its components

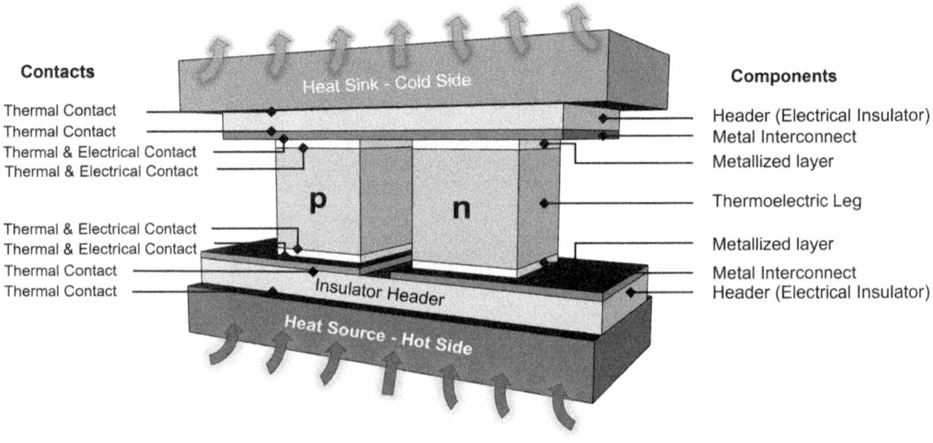

Figure 3.2: Detailed structure of a unicouple thermoelectric generator (TEG). The components include top and bottom headers (or substrate), metal interconnects, metallized layers, and the thermoelectric *p*- and *n*-type legs. Interfaces (contacts) are the source of parasitic resistance in a thermoelectric device.

that can introduce parasitic thermal and electrical contact resistances (Figure 3.2). The overall performance of the TE module is highly dependent on the effective reduction of these resistances.

3.3 Electrical contact resistance

Contact resistance at the interface between TE legs and metal interconnects is one of the most important and challenging issues that must be understood for the effective design of TE devices. Since TE materials are generally semiconductors, achieving an ohmic contact at the junction with a metal interconnect is nearly impossible. In order to minimize the total resistance of the device and ultimately increase its output power and efficiency, a metallized layer is typically deposited on the surface of the TE leg that will be in contact with the solder or brazing material.

The contact resistance of this junction and all the other contacts, i.e., TE leg/solder, solder/metal interconnect, and metal interconnect/header (depicted in Figure 3.2), can be characterized by a scanning probe measurement system on a TE unicouple (Figure 3.3a–c). This system operates on a four-probe measurement configuration, where a constant current is passed through a TE leg while a localized scanning probe measures the voltage profile across it. By scanning between the top and bottom contacts, the local electrical resistance and contact resistances can be determined. Figure 3.3c and d schematically shows the scanning probe measurement setup and its measurement across a real TE leg. Minimizing contact resistance is essential for maximizing the overall performance of the device (Figure 3.3e and f). In general, for a high-performance TE device, the contact resistance should be below 1 μΩ· cm^2. The contribution of contact resistance can be defined using the following equation [70, 71]:

$$(ZT)_d = (zT)_m \left(\frac{L}{L + 2R_c\sigma} \right) \tag{3.1}$$

where $(ZT)_d$ is the device ZT (with uppercase Z), $(zT)_m$ is the material zT (with lowercase z), L is TE leg length, R_c is the contact resistance, and σ is the electrical conductivity of the TE material. According to eq. (3.1), $(ZT)_d$ is a proportion of $(zT)_m$ and is always less than it. Accordingly, in addition to the material zT, parameters such as contact resistance, leg length, and the individual electrical conductivities of the TE legs collectively contribute the overall device performance.

TE devices with minimum contact resistance are required to mitigate the detrimental drop in device performance. Higher contact resistance and higher electrical conductivity both degrade device performance. While increasing leg length can compensate for the parasitic effects of contact resistance, it makes thin films highly disadvantageous, as their contact resistance dominates device performance. Therefore, depending on the anticipated contact resistance, the electrical conductivity and geometry of the TE legs must be optimized to minimize the deteriorative effect of contact resistance. Experimentally, the device contact resistance can be minimized through an appropriate metallization technique, TE leg dimension design, device fabrication process, interconnect electrodes,

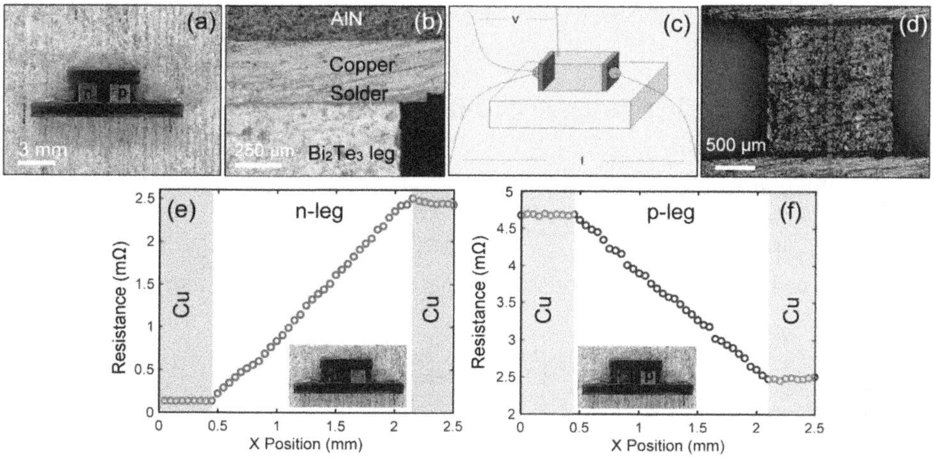

Figure 3.3: (a) Fabricated unicouple thermoelectric device composed of *n*- and *p*-type Bi_2Te_3 alloys. (b) Optical microscopy image showing the junction regions at the AlN/copper/solder/leg interfaces. (c) Schematic diagram of the scanning probe measurement setup used to evaluate device contact resistance. (d) Illustration of the scanning probe measurement locations (red dots) across an actual thermoelectric leg. (e) and (f) Measured contact resistance for the *n*- and *p*-type legs, respectively. Reproduced with permission from Ref. [70]. Copyright © [2019] American Chemical Society.

soldering material, and soldering temperature. The effect of contact resistance is even more detrimental in TE coolers.

3.4 Output power

For wearable applications, the output power of TEGs is more important than their efficiency, as it can be used to continuously power a device or charge a battery. Even, an infinitesimal amount of harvested body energy can change the scenario and enable self-powered devices. The theoretical output power of a TE unicouple without considering the effect of contact resistance, can be calculated using the following equation [72]:

$$P_0 = \frac{[\sigma][\alpha]^2 \Delta T^2 A}{4L} \tag{3.2}$$

where $\Delta T = T_{\mathrm{h}} - T_{\mathrm{c}}$, A is the cross-sectional area, and L is the length of the TE legs. Since the power factor (PF) is equal to $[\sigma][\alpha]^2$, eq. (3.2) can be written as:

$$P_0 = \mathrm{PF} \cdot \frac{A}{4L} \cdot \Delta T^2 \tag{3.3}$$

Equation (3.3) shows the contribution of TE materials' properties (PF), leg dimensions ($A/4L$), and the applied temperature differentiation (ΔT) on the output power of TEGs.

3.5 Thermoelectric device efficiency

The definition of zT for a single material is not physically meaningful because a minimum of two materials are needed to fabricate a TE device, which is typically an array of n- and p-type semiconductor legs. Therefore, the maximum conversion efficiency (η) of TEGs, considering both n- and p-type materials, is calculated using the following equation [58]:

$$\eta = \frac{\Delta T}{T_{\mathrm{h}}} \cdot \frac{\sqrt{1 + ZT_{\mathrm{avg}}} - 1}{\sqrt{1 + ZT_{\mathrm{avg}}} + \frac{T_{\mathrm{c}}}{T_{\mathrm{h}}}} \tag{3.4}$$

where T_{h} is the hot-side temperature and T_{c} is the cold-side temperature, $\Delta T = T_{\mathrm{h}} - T_{\mathrm{c}}$, and ZT_{avg} is the dimensionless figure of merit for the TE device at $T_{\mathrm{avg}} = \frac{T_{\mathrm{h}} + T_{\mathrm{c}}}{2}$. Here, $Z = \frac{\sigma\alpha^2}{\kappa}$, where $[\sigma] = \sigma_p + \sigma_n$, $[\alpha]^2 = (\alpha_p - \alpha_n)^2$, and $[\kappa] = \kappa_p + \kappa_n$. The subscript of p and n shows material properties for p- and n-type legs. Like all thermodynamic engines, the maximum power efficiency is limited to Carnot efficiency ($\Delta T/T_{\mathrm{h}}$). In an ideal condition, when there is no contact resistance and parasitic loss, and both the *n*- and *p*-type material properties are similar and independent of the temperature, $z = Z$.

The efficiency of a TEG can be measured by knowing the total amount of heat applied to one side of the TEG, the power it generates, and the remaining heat that exits from the cold side. Figure 3.4a shows a simple setup for TEG efficiency measurement. This setup includes a heater that provides constant heat for the hot side of the TEG, which in this case is a unicouple (also known as a single-couple) TEG device. It also includes a long, heat-sink-cooled rod, called Q-meter, to dissipate the external heat and maintain the cold side temperature. The heat sink is usually cooled by water circulation. Several thermocouples, placed at specific distances, are embedded inside the Q-meter to measure the temperature gradient along it. In this setup, the output voltage and power of the TEG are measured separately under specific hot and cold side temperatures (Figure 3.4b). The efficiency of the TEG is calculated as follows:

$$\eta = \frac{P_{\text{out}}}{Q_{\text{in}}} = \frac{P_{\text{out}}}{P_{\text{out}} + Q_{\text{out}}} \tag{3.5}$$

where P_{out} is the output power from the TEG, and Q_{in} and Q_{out} are the heat flow from the heat source and the heat flow to the heat sink, respectively. The maximum output power, $P_{\text{out}}^{\text{max}}$, for a device with a known open circuit voltage (V_{oc}) is calculated as follows:

$$P_{\text{out}}^{\text{max}} = \frac{V_{\text{oc}}^2}{4R_{\text{i}}} \tag{3.6}$$

where R_i is the TEG's internal resistance. The Q-meter is utilized to measure Q_{out} using the following equation:

$$Q_{\text{out}} = \kappa A \left(\frac{dT}{dx} \right) \tag{3.7}$$

where κ is the thermal conductivity of the Q-meter (usually made of copper or brass), A is the cross-sectional area of the Q-meter, and dT/dx is the temperature gradient with respect to distance. According to eq. (3.7), Q_{out} can be calculated by knowing the κ of the Q-meter, its cross-sectional area, and the temperature difference between two thermocouples divided by the known distance between them.

The efficiency of a Bi_2Te_3-based TEG device made from state-of-the-art materials for a wearable application is approximately 0.4% at a temperature of 32 °C with a ΔT of 15 °C. Practically, this efficiency would be much lower in an actual wearable device. Although the efficiency of TEGs is low on a wearable platform, their high output power and voltage make them ideal compared to other energy harvesting methods.

According to eq. (3.4), the efficiency of TEGs increases by enhancing both temperature and zT. As shown in Figure 3.4c, the BiTe device efficiency reaches up to 4.2% at 125 °C. Higher efficiency requires discovery of materials with a higher zT. For example, BiTe-based alloys with an average zT of 1.08 for *p*-type and 0.84 for *n*-type demon-

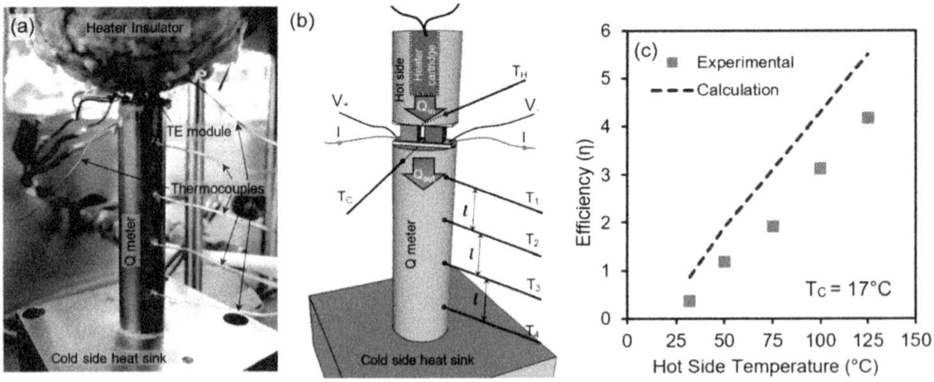

Figure 3.4: (a) Photograph and (b) schematic diagram of the Q-meter setup used to measure TE output power, internal resistance, and efficiency. (c) Measured and calculated BiTe TEG efficiency as a function of the hot-side temperature. Reproduced with permission from Ref. [70]. Copyright © [2019] American Chemical Society.

strated an 8% efficiency at a ΔT of 230 °C [73]. TEG efficiency can also be calculated theoretically using eq. (3.4). The experimental values are always less than theoretical ones, typically due to heat losses from the TEG legs, and/or inaccurate temperature measurements of the TE legs, given the complexity of obtaining the exact temperature at the junctions. Heat losses occur through convection due to the environment and atmospheric circulation, conduction from thermocouples, or radiation, which is more significant at high temperatures.

3.6 Device manufacturer and vendors

While numerous TE companies were founded over the past few decades, most were unsuccessful due to various factors, including a lack of funding, limited governmental support, or the immaturity of the technology itself. In recent years, the expanding applications and technological maturation of TE materials and devices have led to a higher survival rate among new companies in the field. Some of the manufacturers of TE materials, devices, and systems are Coherent (II-VI) Corp (previously Marlow Industries), Laird Technologies, FerroTec, Custom thermoelectric, Tark Thermal Solutions, Align Sourcing, TE Technology Inc., Sheetak, Hi-Z Technology, Phononic, Gentherm, DBK USA, EVERREDtronics, TEC Microsystems, European Thermodynamics Limited, p-n Europe, Adaptive Thermoelectrics, and Wellen Technology.

4 Modeling approaches for body heat harvesting

Understanding the heat flow from the human body core to the skin, and subsequently through the thermoelectric generator (TEG) and to the ambient environment, requires detailed knowledge of heat transport at each interface junction. Multiphysics modeling can simulate these complex conditions, providing a better understanding of the amount of heat rejected from the body and the performance of both thermoelectric (TE) materials and devices. This chapter discusses the modeling approaches for body heat harvesting using TEGs.

4.1 Human body heat and skin temperature

The human (and animal) body generates heat as a byproduct of the metabolic chemical reactions that convert organic matter into operational energy. This heat is distributed by bloodstream and released through the skin by a combination of convection, radiation, and sweat evaporation. The body naturally regulates blood flow in response to changes in ambient temperature by contracting and dilating blood vessels, a process known as physiological vasoconstriction and vasodilation process.

For body energy harvesting, understanding skin temperature is crucial for finding the best spots on the body. These temperatures are highest in areas where blood vessels run close to the skin surface. Body temperature across different locations has been systematically measured [74]. As shown in Figure 4.1a, temperatures are highest on the forehead and gradually decrease down the body to the feet. The ambient temperature also influences this range [75]. For example, in hot environments (47 °C), the difference between the maximum and minimum measured temperatures is only about 1.3 °C; in room temperature (27 °C), the difference is 4.8 °C; and in cold environments (15 °C), the difference is 8.5 °C. Figure 4.1b shows anatomy of skin and different layers under its surface, including epidermis, dermis, and hypodermis, as well as skin and core body temperatures at room temperature.

4.2 TEG design for wearable applications

TEGs can operate under two primary boundary conditions:

(i) Constant temperature: In this condition, the two sides of the device are maintained at fixed, specific temperatures, known as cold (T_C) and hot (T_H) temperatures.

(ii) Constant heat flux: Here, heat is generated at a certain rate, and consequently, neither T_C nor T_H is fixed.

 | https://doi.org/10.1515/9781501514814-004

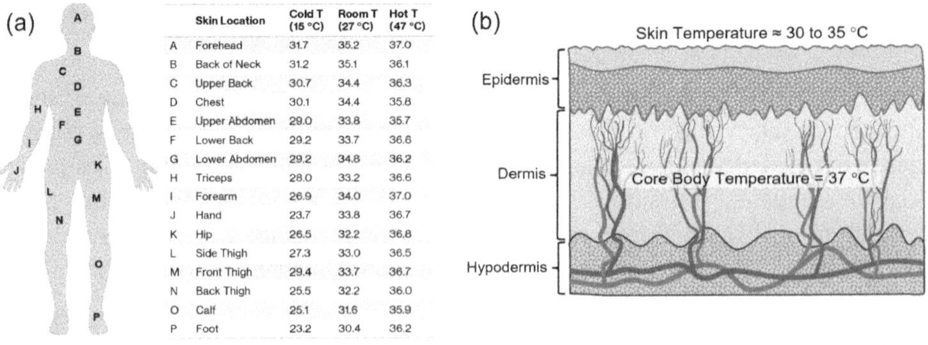

	Skin Location	Cold T (15 °C)	Room T (27 °C)	Hot T (47 °C)
A	Forehead	31.7	35.2	37.0
B	Back of Neck	31.2	35.1	36.1
C	Upper Back	30.7	34.4	36.3
D	Chest	30.1	34.4	35.8
E	Upper Abdomen	29.0	33.8	35.7
F	Lower Back	29.2	33.7	36.6
G	Lower Abdomen	29.2	34.8	36.2
H	Triceps	28.0	33.2	36.6
I	Forearm	26.9	34.0	37.0
J	Hand	23.7	33.8	36.7
K	Hip	26.5	32.2	36.8
L	Side Thigh	27.3	33.0	36.5
M	Front Thigh	29.4	33.7	36.7
N	Back Thigh	25.5	32.2	36.0
O	Calf	25.1	31.6	35.9
P	Foot	23.2	30.4	36.2

Figure 4.1: Human body skin temperature and anatomy. (a) Human body skin temperatures at various ambient temperatures. Note that the greatest difference between maximum and minimum skin temperatures occurs in cold environments. Adapted from Ref. [75] with permission from the Royal Society of Chemistry. (b) Anatomy of human skin, illustrating its topography and the typical variations in core body and skin temperature at room temperature. Reproduced from Ref. [51], with permission from Elsevier.

The optimal properties of TE materials and the device design must be defined based on these specific boundary conditions. Typically, in the literature, most TE devices are characterized under constant temperature boundary conditions where there are no external thermal resistances at the interfaces between the heat source/sink and the device. For this condition, zT is the most important factor; a higher zT indicates a more efficient device. For this reason, most research in the field of TE materials focuses on improving zT. However, this approach is not directly appropriate for applications with constant heat flux where the external thermal resistance exists at the interfaces of the heat source/TEG and the TEG/heat sink. In such cases, this external thermal resistance significantly impacts the heat flow rate through both the interfaces and the TEG itself. When external thermal resistance is approximately zero, the T_H of the TEG is essentially equal to the core temperature of the heat source. A similar scenario applies to T_C. In contrast, when external thermal resistance is high, the heat flow through the TEG is largely unaffected by TEG operation [76].

When a TEG is in direct contact with skin, body heat can pass through the device (Figure 4.2a and b). A consistent heat flow through the TEG can create a temperature gradient (ΔT) between the body skin (hot side) and the ambient environment (cold side), thereby generating an output voltage based on the Seebeck effect.

4.2.1 Parasitic thermal resistances

The successful harvesting of human body heat largely depends on maximizing this temperature gradient (ΔT) across the TEG. However, this is significantly limited by three main external parasitic thermal resistances: (i) Skin as a natural insulator (R_{Skin}): Human body skin naturally acts as an insulator to maintain the core body temperature at 37 °C, thereby impeding the direct transfer of heat from the core of the body to the TEG. (ii) Irregular skin topography ($R_{Contact}$): The irregular surface of the skin creates a secondary thermal barrier to heat flow from the body to the TEG. This occurs primarily because the entire surface of the TEG cannot be in direct contact with the skin, allowing air with low thermal conductivity to fill these microscopic gaps and impede heat transfer. (iii) TEG/ambient air interface ($R_{Heat\ sink}$): Parasitic thermal resistance also exists at the TEG/ambient air interface, primarily due to the lack of ideal convective heat transfer (Figure 4.2b). A significant portion of the available ΔT is lost as a result of these parasitic resistances. Therefore, to enhance the performance of TEGs in wearable applications, it is critical to minimize these resistances through rational design.

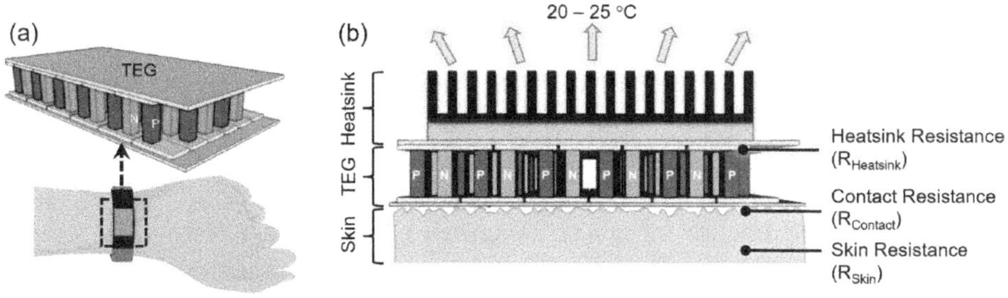

Figure 4.2: Schematic drawing of (a) TEG in contact with the body without a heat sink. (b) Parasitic resistances of skin, skin/TEG, and TEG/heat sink. Reproduced from Ref. [51], with permission from Elsevier.

Heat rejection from the cold side of a TEG to the ambient environment is a critical factor in TE design, typically achieved using heat sinks. There are two primary types of heat sinks: active and passive. Active heat sinks use an external mechanism, such as fans or water cooling, to actively dissipate heat from the cold side, thereby preventing heat buildup. By increasing the convection process, active heat sinks offer high cooling power. Passive heat sinks, on the other hand, have no moving parts and rely on increasing the surface area or heat capacity to dissipate heat into the environment through natural convection and radiation. While passive heat sinks are generally solid state, their cooling power is generally lower than that of active ones.

In wearable applications, active heat sinks are often not viable due to design complexities and their extra power requirements. Traditional passive heat sinks, typically composed of bulky finned structures, aim to maximize heat rejection to the ambient environment. However, several constraints in wearable design, such as body comfort, undesirable size, and aesthetic impact, limit the use of large heat sinks. Thus, an optimal heat sink design is essential, customized to the specific requirements of wearable devices. Such an optimal design would maximize heat dissipation by miniaturizing heat sinks, eliminating the need for separate heat sink entirely, or creatively integrating the device structure itself to function as a heat sink, all without compromising wearability.

4.3 Design and optimization of the TEG parameters

A conventional TEG device is composed of multiple *p*- and *n*-type TE legs that are electrically connected in series while remaining thermally in parallel (Figure 4.2a). More information about the TEGs can be found in Chapter 3. The temperature gradient (ΔT) between the body skin (hot side, T_H) and the ambient temperature (cold side, T_C) generates a steady voltage difference (ΔV). The output voltage and power mainly primarily depend on parasitic thermal resistances; therefore, they can be improved by minimizing thermal resistances and optimizing the TEG design parameters.

The dimensionless TE figure of merit, zT, is a material-dependent property that relies on electrical and thermal conductivity, and the Seebeck coefficient, all of which are functions of temperature. For a TEG device, under fixed temperature conditions, zT remains constant. Hence, the only parameter that can be altered by design is the temperature difference, $\Delta T = T_H - T_C$.

The skin's surface temperature (T_H) is typically a few degrees lower than the core body temperature of 37 °C (Figure 4.1a). Conversely, the ambient temperature (T_C) varies with heat sink and weather conditions. Since T_H is largely determined by skin location and wearability conditions, and thus cannot be significantly altered, T_C emerges as a more practical variable for improving TEG device performance. In a hot environment where the ambient temperature exceeds skin temperature, the positions of the T_H and T_C are reversed; however, the wearable TEG continues to generate a voltage, albeit with inverted polarity.

4.4 Wearable TEG model

The most comprehensive model for calculating the heat transport behavior of a wearable TEG in direct contact with body skin is based on a quasi-three-dimensional approach. This model considers thermal conduction and convection, parasitic thermal effects, Joule heating, and Peltier heating and cooling.

In order to solve this model, a complete electrical circuit acting as the thermal equivalent circuit of and surrounding TE legs was developed (Figure 4.3). Two temperature differentials are defined in the model: the extrinsic temperature difference (ΔT_e) is defined as $\Delta T_e = T_b - T_a$, where T_b and T_a are core body temperature and ambient, respectively; and the intrinsic temperature difference (ΔT_i) is defined as $\Delta T_i = T_h - T_c$, where T_h and T_c are the hot- and cold-side temperatures of the TE legs, respectively.

Upon connecting the TEG to an external load, an electrical current is produced, resulting in Joule heating due to the electrical resistance of the materials. In this model, it is assumed that the Joule heating is equally distributed between the hot and cold sides of the TEG. This phenomenon is represented in the model by two Joule heating sources, $Q_j/2$, located at the top and bottom of each TE leg, where $Q_j = R_{TE} I^2$. In this equation, I is the electrical current, and R_{TE} is the electrical resistance of each TE leg, which depends on its geometry and electrical resistivity. The net effect of the Joule heating is reduction of ΔT_i [51].

Peltier heat (Q_p) is another phenomenon that refers to the heat released or absorbed at the junction of a TE leg and metal interconnect when an electric current (I) passes through them. Peltier heat, described by $Q_p = \Pi I$ (where Π is the Peltier coefficient), is absorbed at the cold side and rejected at the hot side of the TEG. The Peltier coefficient can be calculated from $\Pi = \alpha T$, where α is the Seebeck coefficient and T is the junction temperature (T_c or T_h). The net effect of Peltier heat is to reduce the temperature differential across the TE leg (ΔT_i), thereby decreasing the generated power.

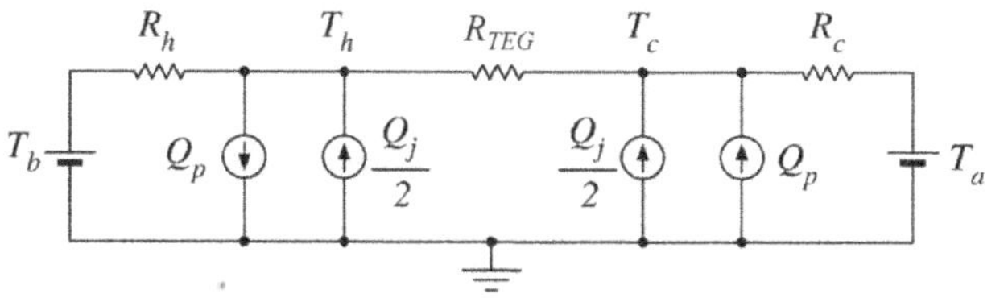

Figure 4.3: The electrical analog represents the thermal equivalent circuit of the model. Reproduced from Ref. [76] with permission from the Royal Society of Chemistry.

The circuit includes the thermal resistance of the TEG legs (R_{TEG}) and the parasitic resistances. The hot-side resistance (R_h) consists of the inherent skin resistance (R_{Skin}), the contact resistance between the skin and the TEG ($R_{Contact}$), and the vertical resistance through the substrate to the leg. The cold-side resistance (R_c) is primarily determined by the heat sink resistance ($R_{Heat\ sink}$) and the vertical resistance of the substrate [75].

Two equations are derived from the analysis of the electrical analog of the thermal equivalent circuit:

$$\frac{T_h - T_c}{R_{TEG}} = \frac{T_b - T_h}{R_h} + \frac{Q_j}{2} - Q_p$$

$$\frac{T_h - T_c}{R_{TEG}} = \frac{T_c - T_a}{R_c} - \frac{Q_j}{2} + Q_p \tag{4.1}$$

The open-circuit voltage (V_{OC}) generated by the TEG is calculated using the following equation:

$$V_{OC} = N\left(\alpha_p + |\alpha_n|\right)\Delta T_i \tag{4.2}$$

where N is the number on p–n leg pairs (equal to half of the total legs), and α_p and α_n are the Seebeck coefficients of p- and n-type legs, respectively. The resistance of the TEG, R_{TEG}, is given by:

$$R_{TEG} = \frac{N}{2}\left(\rho_p + \rho_n\right)\frac{h}{w^2} + R_{ext} \tag{4.3}$$

where h and w are the height and width of each TE leg, and ρ_p and ρ_n are the electrical resistivity of p- and n-type legs, respectively. R_{ext} represents external resistances, including contact resistance at the junction of the TE legs and interconnects as well as the combined resistance of interconnects themselves on both headers.

For an applied load resistance, R_{Load}, the maximum output power of the TEG can be calculated as follows:

$$P_{max} = \frac{V_{OC}^2}{R_{TEG}\left(1 + \frac{R_{Load}}{R_{TEG}}\right)^2} \tag{4.4}$$

4.4.1 Heat leakage around TE legs

The TE legs are connected to two thermally conductive substrates (headers), such as AlN or Al_2O_3. Since the TE legs do not completely fill the space between these substrates, heat flowing into the hot side from the adjacent regions can also contribute to power generation within the device. This heat flows laterally toward and is collected by the TE legs (Figure 4.4a). A similar scenario occurs regarding heat dissipation on the cold side, where heat spreads laterally from the legs to the cold-side substrate, and is then rejected to the ambient by convection. While these lateral heat flows are negligible for thick and highly thermally conductive substrates, they become significant for wearable applications. In such cases, where device size, thickness, and flexibility are crucial, this parasitic effect of lateral heat flow must be considered for accurate calculations and device design.

In addition, a portion of the heat reaching the hot-side substrate conducts through the regions surrounding the TE legs, creating a thermal bypass between the hot and cold substrates. The space between these two substrates can be a vacuum or another medium, such as air or an elastomer. For conventional rigid TEGs, where the area is filled with air, which has a $\kappa \approx$ 0.025 W/m · K at ambient temperature, this bypass effect is expected to be minimal because air is a poor thermal conductor [1]. However, in flexible TEGs, the medium might be an elastomer, such as PDMS, with a $\kappa \approx$ 0.15 W/m · K at ambient temperature. This value is significantly higher than that of air and can, therefore, considerably degrade power generation and overall device performance [75].

4.4.2 Solving equivalent thermal circuit

In this model, an electrical equivalent of the lateral in-plane heat that flows through the substrate, and the out-of-plane thermal bypass of TE legs are calculated using a quasi-three-dimensional model. This model calculates $\Delta T_{\rm i}$ for a given $\Delta T_{\rm e}$ and estimates the TEG power and voltage output. The circuit includes the thermal and electrical resistances of the TE legs, surrounding materials, and parasitic resistances. $R_{{\rm l}i}$ represents the lateral differential thermal resistance of the headers, where $i = 1, 2, \ldots, n$. It is designed for a rectangular ring around each TE leg, estimated by [51, 75]:

$$R_{{\rm l}i} = \frac{1}{\kappa_{\rm h,\,c}} \cdot \frac{dr}{A_{ii}} \tag{4.5}$$

where $\kappa_{\rm h}$ and $\kappa_{\rm c}$ are the thermal conductivity of hot- and cold-side headers, respectively, $A_{ii} = 2\pi r_i t_{\rm h,\,c}$, and dr is the differential cross-plane. Here, r_i is the distance of the center of the rectangular ring from the center of TE leg, $t_{\rm h}$ and $t_{\rm c}$ are the thickness of the hot- and cold-side headers, respectively.

In the cold side, several thermal resistances should be considered. These include the contact resistance between the legs and the header, as well as the thermal resistance between the header and the ambient environment, which is primarily governed by convection. $R_{{\rm c}i}$ represents the sum of the thermal resistances at dr for the cold-side header, calculated as follows:

$$R_{{\rm c}i} = \frac{1}{\kappa_{\rm c}} \cdot \frac{t_{\rm c}}{A_{{\rm c}i}} \tag{4.6}$$

where $A_{{\rm c}i} = 2\pi r_i dr$. The filler between two headers also has thermal resistance. In rigid TEGs, this filler is typically air, whereas in flexible devices, it might be a different material like polydimethylsiloxane (PDMS). Each rectangular ring on the bottom header is connected to a corresponding ring on the top header through a vertical thermal resistance, denoted as $R_{{\rm v}i}$, which is defined as follows:

$$R_{{\rm v}i} = \frac{1}{\kappa_{\rm f}} \cdot \frac{L}{A_{{\rm f}i}} \tag{4.7}$$

where $\kappa_{\rm f}$ is the thermal conductivity of the filler material, L is the thickness of the filler or internal height of the TEG between two headers, and $A_{{\rm f}i} = 2\pi r_i dr_i$ [51].

The network in Figure 4.4b can be solved by defining a repeating block and its corresponding transfer matrix, T [51]:

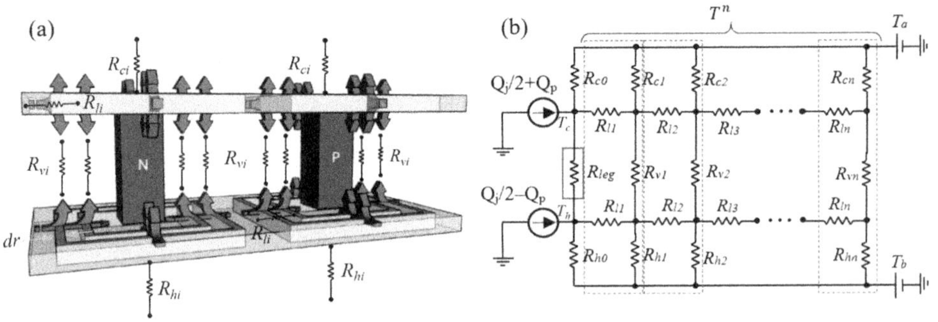

Figure 4.4: (a) Schematic showing the different heat transfer pathways, and (b) the complete thermal circuit model illustrating heat flow within and around the TEG legs. Reproduced from Ref. [51], with permission from Elsevier.

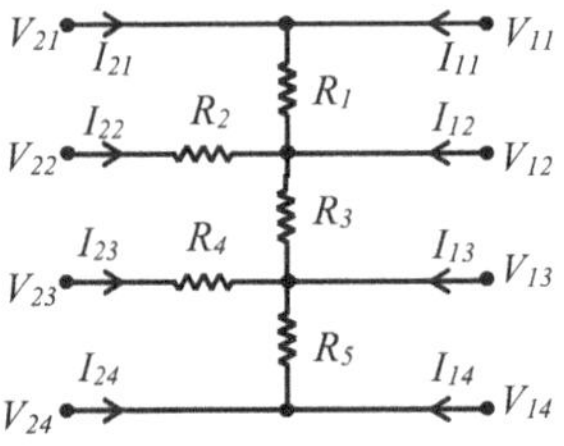

$$\begin{bmatrix} V_{11} \\ V_{12} \\ V_{13} \\ V_{14} \\ I_{11} \\ I_{12} \\ I_{13} \\ I_{14} \end{bmatrix} = T \begin{bmatrix} V_{21} \\ V_{22} \\ V_{23} \\ V_{24} \\ -I_{21} \\ -I_{22} \\ -I_{23} \\ -I_{24} \end{bmatrix}$$

$$T = \begin{bmatrix} 1 & 0 & 0 & 0 & 0 & 0 & 0 & 0 \\ 0 & 1 & 0 & 0 & 0 & R_2 & 0 & 0 \\ 0 & 0 & 1 & 0 & 0 & 0 & R_4 & 0 \\ 0 & 0 & 0 & 1 & 0 & 0 & 0 & 0 \\ \frac{1}{R_1} & -\frac{1}{R_1} & 0 & 0 & 1 & -\frac{R_2}{R_1} & 0 & 0 \\ -\frac{1}{R_1} & \frac{R_1+R_3}{R_1R_3} & -\frac{1}{R_3} & 0 & 0 & \frac{R_1R_2-R_1R_3-R_2R_3}{R_1R_3} & -\frac{R_4}{R_3} & 0 \\ 0 & -\frac{1}{R_3} & \frac{R_3+R_5}{R_3R_5} & -\frac{1}{R_5} & 0 & -\frac{R_2}{R_3} & -\frac{-R_3R_4-R_3R_5-R_4R_5}{R_3R_5} & 0 \\ 0 & 0 & -\frac{1}{R_5} & \frac{1}{R_5} & 0 & 0 & -\frac{R_4}{R_5} & 1 \end{bmatrix}$$

where R_1, R_2, R_3, R_4, and R_5 correspond to R_{ci}, R_{li}, R_{vi}, R_{li}, and R_{hi}, respectively. The resistor network can be replaced by T^n, as shown in Figure 4.4b. The temperature difference across the TE leg is then calculated by solving the simplified equivalent circuit after T^n.

Simpler methods exist for calculating the performance of the wearable TEG, considering the thermal resistances. For example, 1D models are used, though they commonly neglect in-plane heat transfer for simplification. Alternatively, due to symmetry, the equivalent middle temperature of the R_{vi} can be considered a fixed $(T_h + T_c)/2$. This effectively splits the resistor network into only top and bottom sections, conveniently facilitating an equivalent Thevenin circuit model for heat spreading around the TE legs [75].

4.4.3 Core body heat absorption by TEG

The heat flowing from the core body to the TEG encounters two main thermal barriers: (i) Body skin: This is a physiological that varies with age, weight, body fat, and gender. (ii) Contact resistance at the skin/TEG interface: This arises from the roughness of the surfaces. Since skin is soft and deformable, this latter barrier can be minimized by applying external pressure.

The thermal contact resistance (R_{Contact}) at the surface junction of two dissimilar materials, with a contact area of A, can be estimated by a factor known as the thermal contact conductance coefficient or the heat transfer coefficient (h_c), where $R_{\text{Contact}} = 1/h_cA$. The h_c can be calculated as follows:

$$h_c = 1.25\kappa_h \frac{\Delta a}{\sigma}\left(\frac{P}{H_C}\right)^{0.95} \tag{4.8}$$

where κ_h represents the harmonic mean thermal conductivity, Δa denotes the average surface asperity slope, σ is the RMS surface roughness that quantifies the standard deviation of the height values on the skin's surface, P is the applied pressure, and H_C is the microhardness of the skin [75]. The value of h_c varies with pressure and across different locations on the body. A larger heat transfer coefficient corresponds to a smaller contact resistivity. Accordingly, in the human body, the largest h_c values (ranging from 50 to 75 $W/m^2 \cdot K$ for a tight fitting) are found on the forearm and triceps. Surprisingly, the forehead, despite having the highest body temperature, exhibits the smallest h_c (in the range of 25–40 $W/m^2 \cdot K$). Lossec et al. [77] have reported skin heat transfer coefficients ranging from 20 to 100 $W/m^2 \cdot K$, which can be used for calculations.

4.4.4 Heat rejection at the cold-side interface

In order to maintain a constant temperature difference across the TEG and increase its output power, the heat flow entering the TEG from the hot side and exiting from the cold side must be consistent. This necessitates high heat rejection at the cold side of the TEG. As discussed earlier, a heat sink is a common component for passive heat dissipation. For optimal performance, the heat sink should not be covered by the device design or during usage. Heat sink performance can be further improved by air flow or enhanced convection resulting from natural body motion. Therefore, naturally more mobile locations, such as the hand, are ideal. Nonetheless, the design of heat sinks for wearable electronics requires avoiding bulky structures, even though a smaller heat sink surface area inherently results in a higher thermal resistance on the cold side.

There are different methods to calculate the thermal resistance of heat sinks. For example, for a heat sink with m number of fins, $R_{\text{Heat sink}}$ can be calculated from heat conduction principles using the following equation [78]:

$$R_{\text{Heat sink}} = \frac{1}{\eta_0 h(mA_f + A_b)} \tag{4.9}$$

where η_0 is the overall efficiency of the fin array, h is the heat transfer coefficient of the medium around the fins, A_f is the surface area of a single fin, and A_b is the exposed area of its base plate. For instance, given $\eta_0 = 95\%$ and $h_{air} = 5\ \text{W/m}^2 \cdot \text{K}$, the effective heat transfer coefficient of an aluminum heat sink with fin dimensions of 10 mm × 10 mm × 0.5 mm equals 100 $\text{W/m}^2 \cdot \text{K}$. Evidently, the thermal conductivity of the heat sink material and the air velocity blowing across the heat sink must also be considered.

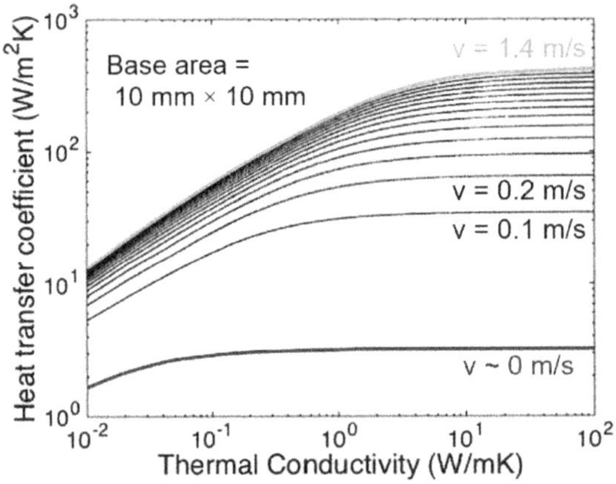

Figure 4.5: Variation of the heat sink heat transfer coefficient with material thermal conductivity and airflow velocity. Adapted from Ref. [75] with permission from the Royal Society of Chemistry.

Another model, which assumes forced convection as the main heat transfer mechanism while also considering heat conduction through the heat sink fins, has been suggested [79]. In this model, for a given air velocity and heat sink geometry, the Nusselt number, which is a dimensionless quantity representing the ratio of convective to conductive heat transfer across a heat sink, is calculated. This value can then be used to estimate the thermal resistance of the heat sink via the equation $R_{\text{Heat sink}} = \frac{1}{Ah_{\text{Heat sink}}}$, where A is the area of the TEG and $h_{\text{Heat sink}}$ is the heat transfer coefficient of the heat sink [75].

Figure 4.5 illustrates the calculated heat transfer coefficient for a 10 mm × 10 mm heat sink with 11 fins (0.5 mm thick and 5 mm high) as a function of the heat sink material's thermal conductivity, for various air velocities. These velocities range from zero for stationary conditions to 1.4 m/s for normal walking speed. According to Figure 4.5, the heat transfer coefficient is a strong function of convective cooling, indicating that higher power can be achieved from wearable devices when there is airflow over the heat sink or when the device itself is moving. Therefore, body locations that naturally exhibit more motion, such as the wrist and forearm, are ideal for self-powered wearables. Furthermore, at higher air velocities, the incremental benefit of air convection weakens. This suggests that the performance of the TEG does not improve proportionally when a person runs at a faster speed. This phenomenon occurs because, at high air velocities, the convective heat transfer coefficient increases more slowly relative to velocity, and thermal resistance is no longer dominated by convection.

According to Figure 4.5, the heat transfer coefficient saturates at a certain thermal conductivity of the heat sink material. As the air velocity increases, this saturation point shifts to higher thermal conductivity. Interestingly, at a convective air velocity of 1.4 m/s, there is no practical improvement gained by using a material with a thermal conductivity greater than 10 W/m · K. While heat sinks are typically made of aluminum (which has a thermal conductivity ranging from 150 to 230 W/m · K), for small heat sinks with short fin heights (which are ideal for wearable devices), heat does not need to travel a large distance to be dissipated. Therefore, various materials with a thermal conductivity below 10 W/m · K are practical. This means flexible materials like conductive polymers, with thermal conductivities around 5 W/m · K, can be used for appropriate wearable heat sink designs. $h_{\text{Heat sink}}$ can range from 5 W/m^2 · K for natural air convection to 150 W/m^2 · K for forced air convection. Typical materials and device parameters at room temperature (27 ° C) that can be used in these calculations are listed in Table 4.1.

Table 4.1: Thermoelectric materials and device parameters at room temperature (27 °C) used in calculations.

Parameter	Value *
Skin temperature	34 °C
Substrate thickness	0.5 mm
Substrate thermal conductivity	200 W/m · K
Copper interconnect resistivity	10^{-8}Ω · m
Copper interconnect thickness	0.02 mm
Electrical contact resistivity	10^{-7}–10^{-6} · cm^2
Electrical conductivity of *p*-type BiTe leg	1,000–1,250 S/cm
Seebeck coefficient of *p*-type BiTe leg	200 μV/K
Thermal conductivity of *p*-type BiTe leg	1.5 W/m · K
Electrical conductivity of *n*-type BiTe leg	875 S/cm
Seebeck coefficient of *n*-type BiTe leg	200 μV/K
Thermal conductivity of *n*-type BiTe leg	1.5 W/m · K
Thermoelectric leg height	1–4 mm
Thermoelectric leg base area	0.25–4 mm^2

Table 4.1 (continued)

Parameter	Value *
TEG area	100–256 mm^2
Thermal conductivity of air	0.025 W/m · K
Heat transfer coefficient of skin	20–100 $W/m^2 \cdot K$
Heat transfer coefficient of heat sink	5–150 $W/m^2 \cdot K$

*These are example values and may vary depending on materials properties and device characteristics.

4.4.5 Load resistance optimization for maximum power

TEG is a nonlinear device where Joule heating and Peltier heating/cooling effects actively change the output power. Consequently, it does not exactly follow the maximum power transfer theorem (eq. (3.6)), which states that the maximum power is given by $P_{max} = \frac{V_{OC}^2}{4R_{Load}}$, where R_{Load} is the external load resistance. Figure 4.6a demonstrates the effect of Joule and Peltier heating relative to the total heat flowing through the TEG. The contribution of the Joule heating is negligible since the amount of electrical current passing through the TEG is small; accordingly, its effect is neglected in this model. However, the Peltier effect is critical. As shown in Figure 4.6b, when the Peltier effect is considered, there is a noticeable drop in the output power of TEG. Furthermore, in this case, the maximum power no longer occurs when $R_{Load} = R_{TEG}$; it shifts to a higher value, which is consistent with the analytical expression as $R_{Load} = R_{TEG}\sqrt{1+ZT}$, reported in the literature [80].

Figure 4.7 shows the temperature profile of the forearm from the core body to the ambient environment at two distinct temperatures of 27 °C and 15 °C. It reveals that although the core body temperature is 37 °C, only a small portion of the external

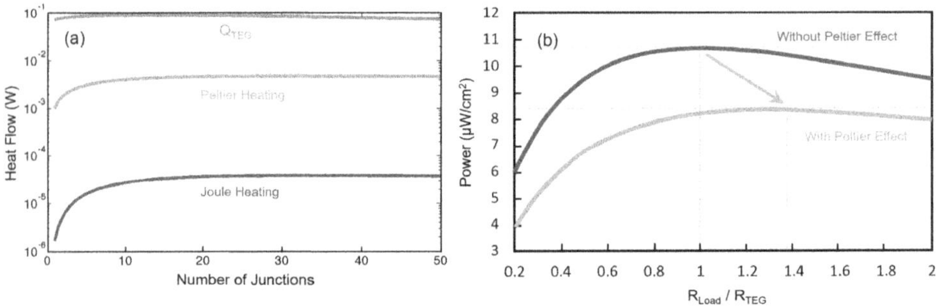

Figure 4.6: (a) Comparison of heat flow from the TEG and heat losses from Peltier and Joule heating versus the number of junctions. (b) Output power as a function of the R_{Load}/R_{TEG} ratio, with and without the Peltier effect. Adapted from Ref. [75] with permission from the Royal Society of Chemistry.

ΔT (< 1 °C) drops across the TE leg. The ΔT across the TE leg changes from 0.5 °C at an ambient temperature of 27 °C to 0.9 °C at 15 °C, indicating a higher ΔT at lower ambient temperatures. However, the percentage of the total available ΔT across the TE leg, relative to the overall temperature difference between the core body and the ambient environment, remains constant at approximately 4% to 5%.

As Figure 4.7 illustrates, approximately 50–60% of the total temperature drop is caused by internal skin resistance, with the remaining portion attributed to the TEG and the heat sink. Skin resistance is an inherent physiological factor, and from an engineering design perspective, one can only select a body location that exhibits the lowest possible skin resistance. Therefore, the primary focus of the engineering design must be on reducing the thermal contact resistance ($R_{Contact}$) and the heat sink resistance ($R_{Heat\ sink}$). As discussed previously, external pressure can reduce the $R_{Contact}$, but it may cause long-term physical discomfort. $R_{Heat\ sink}$ can be minimized through appropriate heat sink geometry, material selection, and fin optimization; however, this is often limited by other factors such as the overall shape and size of the device.

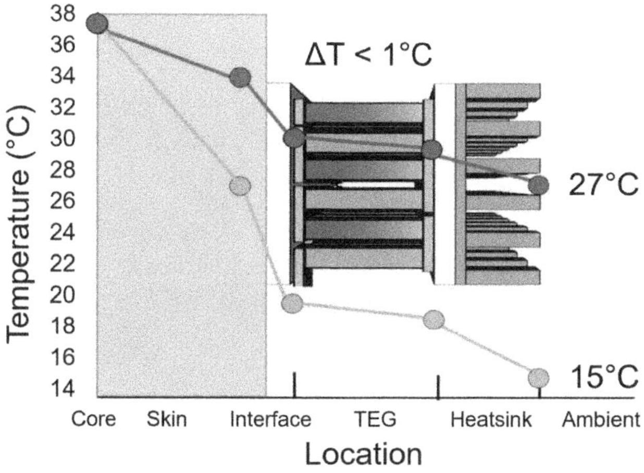

Figure 4.7: Temperature profile from the core body to the ambient air for two ambient air temperatures: 15 °C and 27 °C. A large portion of the temperature difference is lost through the skin, with less than 1 °C passing through the TE legs. Reproduced from Ref. [75] with permission from the Royal Society of Chemistry.

For wearable applications, a higher ΔT and a larger thermal resistance across the TEG are needed. The thermal resistance of the TEG (R_{TEG}) can be expressed as follows:

$$R_{TEG} = \frac{1}{\kappa}\left(\frac{h_{Leg}}{A_{TEG}\,FF}\right) \tag{4.10}$$

where κ is the thermal conductivity of the TE materials, h_{Leg} is the TE leg height, A_{TEG} is the TEG base area, and FF is the fill fraction (FF) of the TE legs. The fill fraction is defined as the ratio of the area occupied by the TE legs to the total TEG base area. Figure 4.8

demonstrates two distinct TEGs with FFs of 12% and 36%. An FF of 12% is considered as a low FF and an FF of 36% is considered a medium to high FF. According to eq. (4.10), R_{TEG} depends on a combination of materials properties (κ) and material and device geometry and design (h_{Leg}, A_{TEG}, and FF). Accordingly, R_{TEG} increases with lower thermal conductivity of the TE materials, longer (taller) TE legs, a smaller TEG base area, and a lower FF of the TE legs.

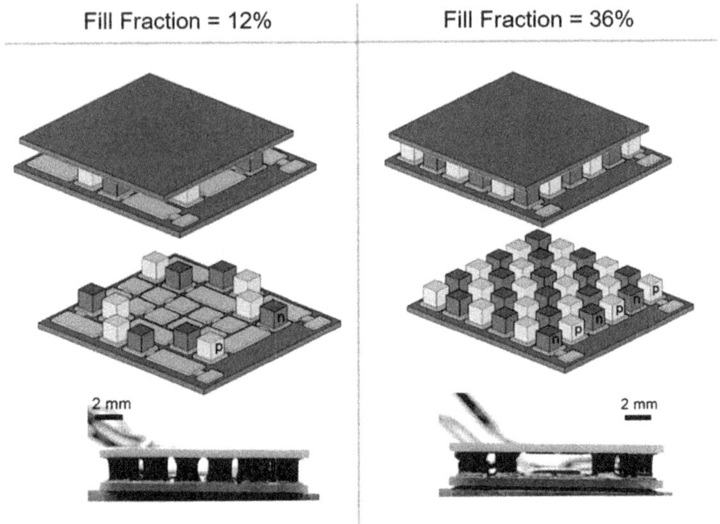

Figure 4.8: Schematic drawings and photographs of the fabricated TEG devices, fabricated with commercial *p*- and *n*-type Bi_2Te_3 legs and demonstrate fill fractions (FFs) of 12% and 36%. Reproduced with permission from Ref. [70]. Copyright © [2019] American Chemical Society.

4.5 Material property requirements

Most research in the field of TE materials for power generation and cooling applications focuses on improving the zT. In cases where power output or cooling power is more critical than efficiency (η) or the coefficient of performance (COP), the power factor ($PF = \sigma\alpha^2$) becomes particularly important [81]. However, specific scenarios may present varying requirements.

Comprehensive model calculations reveal that for body heat harvesting, lower thermal conductivity (κ) is often more crucial than a higher zT. Figure 4.9 compares the output power density versus κ and the Seebeck coefficient (α) for two different zT values: 0.8 (representing an average TE material) and 2.0 (representing an excellent, albeit hypothetical, TE material). Figure 4.9a illustrates this comparison for an ideal heat source and heat sink with a thermal resistance of 1,000 $W/m^2 \cdot K$, which signifies minimal parasitic resistances at the junctions. Figure 4.9b, conversely, compares the two zT values when there are large parasitic resistances on both sides of the TEG, similar to the conditions encountered in body heat harvesting. In this specific case, the heat transfer coefficient at the skin/TEG and TEG/heat sink junctions are 50 and 100 $W/m^2 \cdot K$, respectively.

In both scenarios (Figure 4.9a and b), the output power increases slightly with an increasing α. For a constant zT and κ, a higher α corresponds to a lower electrical conductivity (σ). As the σ decreases, the current, and consequently Joule and Peltier heating, reduce, resulting in higher power output [75].

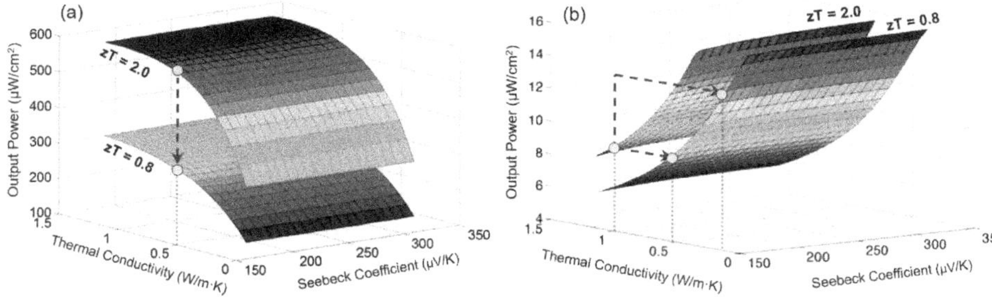

Figure 4.9: Comparison of the output power density versus κ and α for two different zT values. (a) Low thermal resistance at the junction of the TEG headers with the heat source and sink, a condition commonly assumed in the design of TE materials for various applications. (b) High parasitic contact resistance at the junction of TEG headers with the heat source and sink, a condition mainly appropriate for wearable applications. The calculations were performed for TE materials with a 1-mm leg height, a 0.25-mm^2 base area, and a fill fraction (FF) of 25%. Reproduced from Ref. [75] with permission from the Royal Society of Chemistry.

The calculations were conducted assuming a constant FF of 25%, a leg length of 1 mm, and a TE leg cross-sectional area of 0.25 mm^2. both κ and α were varied, while σ was adjusted to maintain a constant zT. Two main scenarios were evaluated: (i) the conventional case of a TEG with low thermal resistances at the cold and hot sides, and (ii) a wearable TEG with high parasitic thermal resistances on both sides [75].

According to Figure 4.9a, a higher zT always results in greater output power for any combination of α, σ, and κ. For both zT values, lowering κ at a given α reduces the output power of the device. This is primarily because σ reduces proportionally with κ to maintain a constant zT, leading to a lower PF. Thus, for small parasitic resistances, a lower PF translates to reduced output power [75].

In contrast, under conditions of high thermal resistance (Figure 4.9b), a lower thermal conductivity (and consequently a smaller PF) leads to an increased output power. This occurs because the parasitic resistances are significantly larger than the TE leg resistances, meaning that only a fraction of the total temperature difference occurs across the TEG itself. Therefore, since the output voltage of a TE leg depends on its Seebeck coefficient and the temperature difference across it (i.e., $V_{OC}=\alpha\Delta T_i$), from eqs. (3.6) and (4.4), the maximum output power (P_{max}) can be expressed as $P_{max} = \frac{V_{OC}^2}{4R_{TEG}} = \frac{(\alpha \Delta T_i)^2}{4R_{TEG}}$. Therefore, $P_{max} \propto \Delta T_i^2$.

Figure 4.9b shows that a material with $\kappa = 0.5$ W/m · K and $zT = 0.8$ can generate the same power as a material with $\kappa = 1$ W/m · K and $zT = 2$. More interestingly, a material with $\kappa = 0.2$ W/m · K and $zT = 0.8$ can generate almost twice the power of a material with $\kappa = 1$ W/m · K and $zT = 2$. This result confirms that for wearable applications, at a TE material level, while a high zT is still important, a lower κ is far more critical than any other parameter [51, 75, 82]. Therefore, polymers, and other printable and flexible TE materials with very low κ are promising candidates for wearable devices.

It should be noted that for both cases, the system requires a high Seebeck coefficient to provide sufficient voltage. Hence, a higher Seebeck coefficient is always desirable.

4.6 Thermoelectric device design for wearables

One of the primary goals in designing TE devices for wearable applications is to optimize the TEG thermal resistance relative to the parasitic thermal resistances. One way to boost the thermal resistance of a TEGs is to reduce the FF. This can be realized by decreasing the number of TE legs or their cross-sectional area. Since the κ of the TE materials is much higher than that of air, reducing the number of legs lessens the thermal bypass between the top and bottom headers, which in turn increases the overall thermal resistance of the module.

Figure 4.10 compares the internal temperature difference (ΔT_i), output voltage, and power density versus FF for three different TEGs, with material thermal conductivities of 0.2, 0.5, and 1.5 W/m · K. Other parameters like leg height and base area are kept constant. ΔT_i decreases as FF increases, i.e., as more TE legs are added (Figure 4.10a). The module with lower thermal conductivity consistently maintains a higher ΔT_i across all FFs. For a material with $\kappa = 1.5$ W/m · K, ΔT_i drops below 1 °C when FF > 15%. This scenario is typical for most commercially available Bi_2Te_3 materials. The TEG with the lowest thermal conductivity generates a higher output voltage across all FFs, and the voltage clearly increases with FF (Figure 4.10b). In terms of output power, the TEG with $\kappa =$ 1.5 W/m · K reaches a maximum power at FF of 3%, after which the power drops at higher FFs (Figure 4.10c). However, the output power of the TEG with $\kappa = 0.2$ W/m · K increases with the FF and reaches a maximum at FF > 25% [75].

A similar behavior is observed when the leg base area changes from 1 mm^2 to 0.0625 mm^2 (Figure 4.10d–f). By reducing the leg base area, ΔT_i and the output voltage increase for the same number of legs. The output power reaches nearly the same maximum value but at a different number of legs. When the leg base area decreases, the maximum power shifts to a higher number of legs. It is worth noting that for smaller TE leg base areas, the TE legs must possess robust mechanical properties to be diced into such small dimensions.

Figure 4.10g–i demonstrates the effect of TE leg height on TEG performance versus FF. Device performance steadily improves with increasing leg height. At a constant FF, ΔT_{i} can increase up to four times for a 4-mm leg compared to a 0.5-mm leg (which is a typical leg length for standard TE cooling devices [83]). As leg length increases, the leg's thermal resistance increases, which helps to maintain high ΔT_{i}. The output voltage and maximum power both increase with leg height, and their picks shift to higher FFs.

4.6.1 Impact of filler material thermal conductivity

There is a substantial area between the two TEG substrates that is not filled with TE materials. More precisely, (100 - FF)% of the area remains open. For instance, in a TEG device with a 25% FF, 75% of the area between the two substrates remains open. This unoccupied space can allow heat to bypass directly from the bottom to the top (hot to the cold) substrate. Typically, this area is filled with air, which has a low thermal conductivity of ~ 0.025 W/m · K. However, this space can also be filled with a secondary medium that has higher thermal conductivity than air. This is especially relevant for flexible devices, as air can be replaced with a polymer. Figure 4.11a compares the output power density versus the FF for four different media filling this space: vacuum with $\kappa \approx 0$, air with $\kappa \approx 0.025$ W/m · K, PDMS with $\kappa \approx 0.15$ W/m · K, and a hypothetical composite comprising 50% air and 50% PDMS with $\kappa \approx 0.08$ W/m · K [75].

All media exhibit similar behavior, reaching a maximum power density at FF < 5% before decreasing, due to the reduced thermal resistance of the TEG as more legs are added. While at low FFs, the medium with the lower thermal conductivity dramatically outperforms the others, above an FF of 20%, the power densities converge to nearly the same values. Thus, the effect of the filling medium's thermal conductivity diminishes at higher FFs. As the thermal conductivity of the medium increases, the maximum power decreases.

PDMS is one of the most commonly used polymers in flexible electronics. According to Figure 4.11a, it can be concluded that flexible TEGs can be fabricated with performance comparable to that of standard rigid TEGs employing air as the gap medium. More interestingly, flexible TEGs with a low FF and an air/PDMS medium can generate up to twice the power of typical commercial rigid TEGs with FFs greater than 25%. Therefore, the analysis of TEG output power versus FF across different media confirms the capability of flexible TEGs to offer better performance and flexibility than typical commercial rigid TEGs, as well as lower overall fabrication costs.

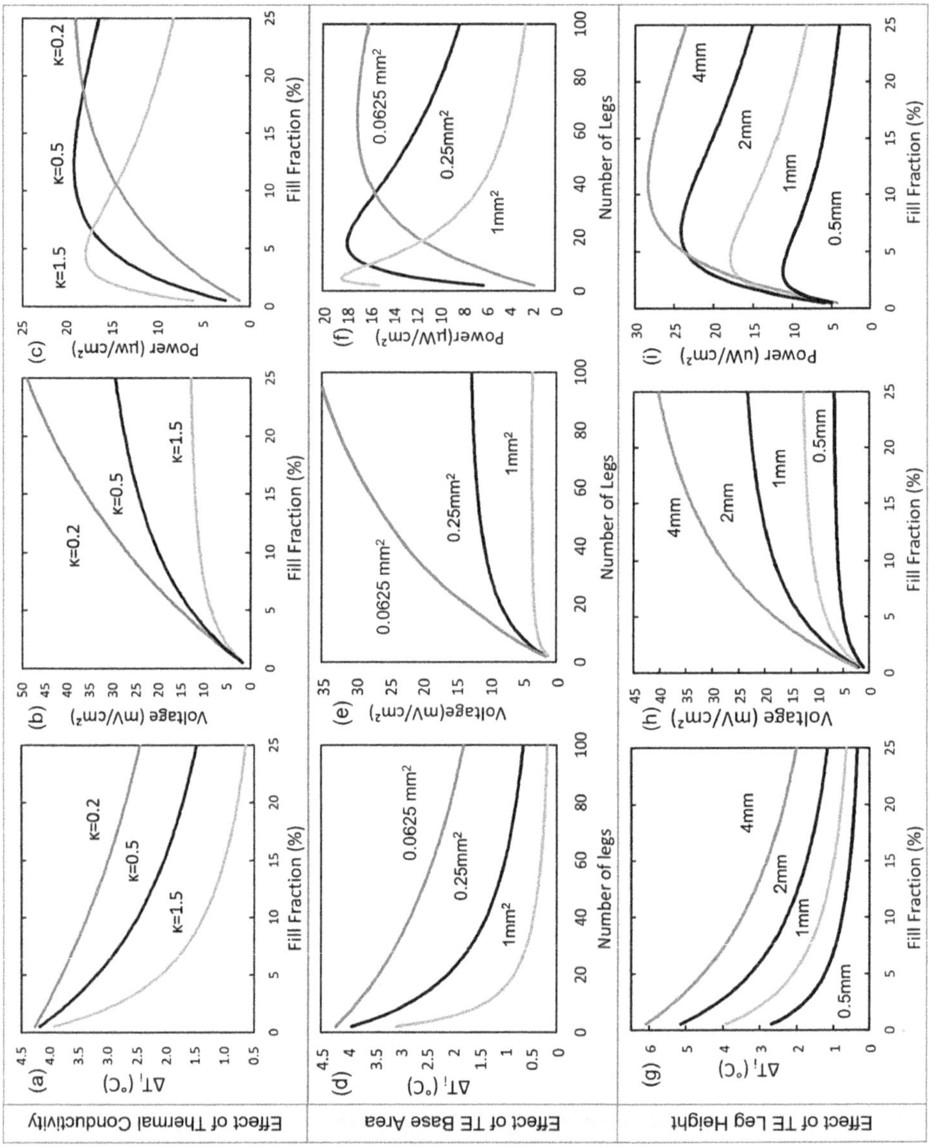

Figure 4.10: Temperature difference, output voltage, and power density of three different TEG configurations. (a–c) Effect of TE thermal conductivity: Varying thermal conductivity versus the FF. The thermal conductivities of the TE materials are 0.2, 0.5, and 1.5 W/m · K. (d–f) Effect of TE base area: Temperature difference, output voltage, and power density of three different TEGs with varying TE base areas versus the number of legs. The base areas of the TE materials are 1, 0.25, and 0.0625 mm^2. (g–i) Effect of TE leg height: Temperature difference, output voltage, and power density for four different TEGs with varying TE leg lengths versus FF. The leg lengths of the TE materials are 0.5, 1, 2, and 4 mm. Adapted from Ref. [75] with permission from the Royal Society of Chemistry.

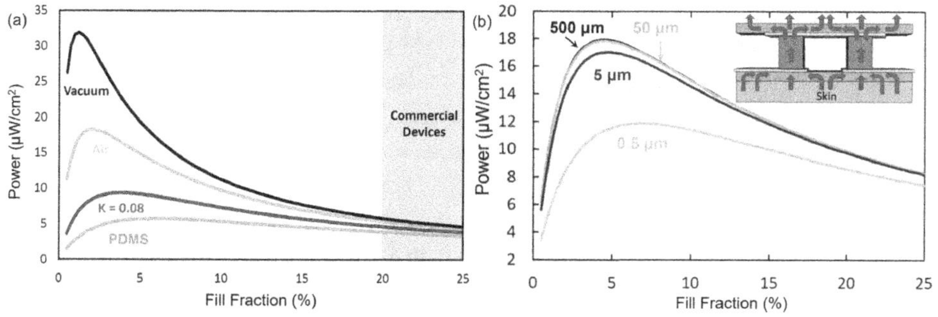

Figure 4.11: Power density versus the fill fraction for: (a) Different filling media: including vacuum, air ($\kappa \approx 0.025$ W/m · K), a composite of PDMS and air ($\kappa = 0.08$ W/m · K), and PDMS ($\kappa \approx 0.15$ W/m · K); and (b) Various substrate thicknesses. The performance of substrates thicker than 5 μm is nearly identical. The inset shows the heat flow path from the bottom (hot) substrate (in contact with the skin) to the ambient. Adapted from Ref. [75] with permission from the Royal Society of Chemistry.

4.6.2 Impact of substrate thickness

The absorbed or rejected heat flows laterally through the substrates. On the hot side, heat from the skin travels to the substrate, which is typically made of a highly conductive ceramic like aluminum nitride (AlN) or alumina (Al_2O_3), flows laterally across it, and then enters the TE legs. On the cold side, heat flows from the legs to the top substrate, disperses laterally, and is eventually dissipated to the ambient. The thickness and thermal conductivity of the substrates dictate this lateral thermal conduction.

Figure 4.11b shows the effect of substrate thickness on the TEG output power density versus the FF. For all substrate thicknesses, the power reaches a maximum (at an FF below 8%) and then decreases as the FF increases. At a low FF (<15%), the power increases as the thickness increases from 0.5 μm to 500 μm; however, the difference in performance above 5 μm is negligible. At a high FF (>20%), the variation in power generation is minor. It can thus be concluded that the effect of substrate thickness is minimal. Therefore, for flexible TEGs, thinner substrates can be used at high leg FFs without significantly degrading device performance.

4.6.3 Impact of heat spreaders

Heat spreaders are thin, thermally conductive materials or structures designed to evenly distribute heat from a localized source over a larger surface area. They serve as solutions for addressing parasitic thermal resistance, extending beyond the intrinsic capabilities of the TEG to absorb or reject heat. Their base area is generally larger than that of the TEG, and they are typically fabricated from materials with superior thermal conductivity and, in the case of flexible TEGs, good flexibility. TEGs often have specific header sizes that are limited by design and cost constraints. On the hot side, heat spreaders can be positioned between the skin and the TEG, absorbing heat from the surrounding skin and distributing it laterally to maintain a constant temperature at the junction with the TEG. On the cold side, the spreader behaves similarly to a heat sink, aiding in the rejection of heat across a larger surface area. In scenarios where heat sink usage is limited, heat can be rejected further away from the cold-side header, helping to delocalize heat from the TEG header itself. Copper is a common choice for heat spreaders due to its high thermal conductivity of ~385 W/m · K and its mechanical strength, making it suitable for flexible, wearable applications. Other popular heat spreader materials include aluminum and carbon, which are typically used in the form of thin plates or foils.

Figure 4.12a shows the output power of a TEG with a 1-cm^2 substrate area versus its FF, for various heat spreader sizes [75]. The output power increases as the heat spreader size grows, indicating that external heat spreaders can significantly contribute to improving TEG power. However, the performance improvement diminishes when the heat spreader size becomes too large relative to the TEG, as heat then needs to travel a practically limited long distance. The effect of heat spreader thickness mirrors that of substrate thickness (Figure 4.11b), where thicker spreaders contribute to higher output power. Figure 4.12b shows how the spreader can be positioned on a flexible TEG device. Heat spreaders can be integrated into wearable TEG devices to enhance output power with minimal additional cost and without increasing the volume of the TE materials or devices. In a wearable device, its band could even be utilized as a heat spreader to absorb and reject heat over a larger surface area.

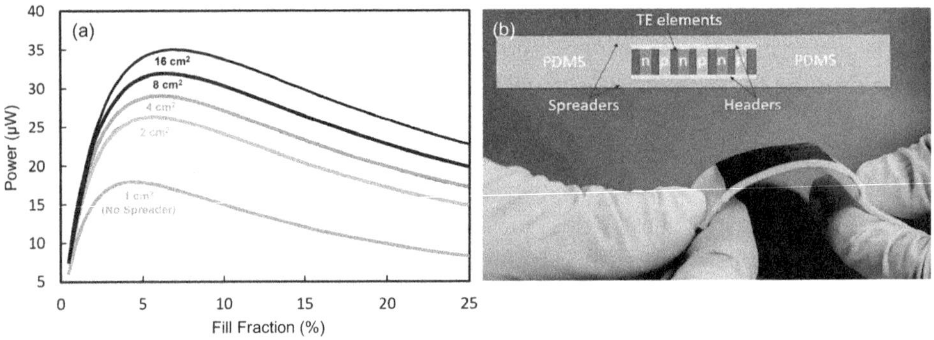

Figure 4.12: (a) Output power versus the fill fraction for a 1-cm^2 TEG substrate area under various heat spreader sizes. Adapted from Ref. [75] with permission from the Royal Society of Chemistry. (b) Schematic and photograph of a flexible thermoelectric device, illustrating heat spreaders covering the full top and bottom surfaces of the TEG. Reproduced from Ref. [1], with permission from Elsevier.

4.6.4 Designing factors of TEGs

The rational design of wearable TEGs requires the optimization of several factors, including leg geometry, contact resistance, interconnects, the fill fraction (FF), and the operating temperature, as well as dimensionless design factors for thermal and electrical conduction. The resulting optimal power and efficiency can be expressed by considering these design factors [84]. For identical TE properties in the *p*- and *n*-type legs, and assuming that these properties are independent of temperature, P_{max} can be calculated using the following equation [85, 86]:

$$P_{max} = \frac{\alpha^2 \; A \cdot N \; (T_H - T_C)^2}{2\rho \; (\gamma + l) \left(1 + 2\lambda \frac{l_c}{l}\right)^2} \tag{4.11}$$

where α is the Seebeck coefficient, ρ is the electrical resistivity, A is the total cross-sectional area the TE legs, N is the number of p–n pairs, T_H and T_C are the temperatures of the hot and cold sides of the TEG, respectively. The parameter $\gamma = \frac{2\rho_c}{\rho}$, where ρ_c is the specific contact resistivity and l is the length (height) of the TE leg. The parameter $\lambda = \frac{\kappa}{\kappa_c}$, where κ is the thermal conductivity of the TE material and κ_c is the effective thermal conductivity of the contact layer, and l_c is the thickness of the contact layer. Using the same parameters, the efficiency (η) of the TEGs can be calculated using the following equation:

$$\eta = \eta_{Carnot} \left[\left(1 + 2\lambda \frac{l_c}{l}\right)^2 \left\{ 2 - \frac{1}{2}\eta_{Carnot} + \left(\frac{4}{Z_m T_H}\right) \left(\frac{l + \gamma}{l + 2\lambda l_c}\right) \right\} \right]^{-1} \tag{4.12}$$

where η_{Carnot} is the Carnot efficiency (i.e. $(T_H - T_C)/T_H$) and Z_m is the modified TE figure of merit of the module, equal to $\frac{(\alpha_p - \alpha_n)^2}{R_{TEG} \cdot K_{TEG}}$, in which α_p and α_n are the Seebeck coefficients of p- and n-type legs, R_{TEG} is the electrical resistance of a p–n pair and K_{TEG} is the thermal conductance of a p–n pair [87], where

$$R_{TEG} = R_{Legs} + R_{Contact} = \frac{\rho_p l}{A_p} + \frac{\rho_n l}{A_n} + R_{Contact} \tag{4.13}$$

where R_{Legs} is the total resistance of the *p*- and *n*-type legs and $R_{Contact}$ is the total contact resistance of the legs. K_{TEG} is calculated using the following equation:

$$K_{TEG} = \frac{\kappa_p l}{A_p} + \frac{\kappa_n l}{A_n} \tag{4.14}$$

According to these equations, larger γ increases the electrical resistance of the TEG, while larger λ reduces the ΔT across the TEG resulting in a lower output voltage. In general, larger values of γ and λ results in a reduction of P_{max} and η in a wearable platform [88, 89].

5 Required components for self-powered wearable device

A wearable thermoelectric (TE) platform must be designed and optimized at the material, device, and system levels to achieve maximum performance and a desirable appearance. At the material level, a high figure of merit (zT) and low thermal conductivity (κ) are necessary to enhance output power and the temperature difference (ΔT) across the thermoelectric generator (TEG). At the device level, mechanical robustness and an appropriate size are needed for durability and aesthetics, while a high Seebeck coefficient (α) is required to maximize voltage output and the efficiency of the DC–DC boost converter. At the system level, the TEG must operate within a power management unit (PMU) that has its own specific requirements for maximum efficiency. Furthermore, since TEGs can continuously harvest energy, some of this power can be stored in a battery for later use. For example, when a person is sleeping or inactive, the device can save most of the harvested energy for use during extensive physical activity. In this chapter, the essential components of a self-powered wearable device are discussed, and the appropriate systems are introduced and analyzed.

5.1 Thermoelectric materials for wearables

Since wearables typically operate around room temperature and rarely exceed 50 °C, the ideal TE materials are those that perform best at lower temperatures. Currently, bismuth telluride (BiTe or Bi_2Te_3)-based alloys are the best candidates for this purpose. The highest-performing devices use these alloys, and they will likely remain the standard until a reliable substitute is discovered. BiTe alloys have undergone extensive development over several decades.

Figure 5.1 compares the Seebeck coefficient, thermal conductivity, and zT among state-of-the-art p-type $(Bi_xSb_{1-x})_2Te_3$ [51, 73, 90–95] and n-type $Bi_2Te_{3-x}Se_x$ materials [1, 95–103]. While some materials exhibit a large zT value ($zT > 1$) above 50 °C, they are not suitable for body heat harvesting because their Seebeck coefficient is low and their thermal conductivity is high at body temperature. Overall, n-type materials underperform compared to p-type alloys and have a lower zT value. Although there is a need to improve the zT and decrease the thermal conductivity of Bi_2Te_3 alloys (with a particular focus on n-type materials), it appears that the current alloys have reached their maximum possible performance, and a fundamental innovation would be required to significantly enhance their properties.

Based on Figure 5.1a–c, p-type materials exhibit a broad range of properties. The room temperature Seebeck coefficient for these materials ranges from 180 to 260 μV/K, with thermal conductivity ranging from 0.7 to 1.4 W/m · K. The zT values for these materials are between 0.75 and 1.75. In contrast, the n-type Bi_2Te_3 materials show differ-

 | https://doi.org/10.1515/9781501514814-005

ent behavior at room temperature (Figure 5.1d–f). Their Seebeck coefficient has a wider range, from −160 to −310 µV/K. The thermal conductivity is similar to *p*-type materials, ranging from 0.75 to 1.6 W/m · K. However, the *zT* values are much more consistent, staying close to a range of 0.75 to 1.1.

BiTe-based TE materials have been widely used in commercial modules for cooling and low-grade power generation. Historically, the main challenge in TE materials has been improving the *zT*. To enhance efficiency at body temperature, more experimental research is needed to achieve $zT > 2$ at ambient temperature for both *p*- and *n*-type materials. While some alloys currently show $zT > 1$, they still do not fully meet the requirements for wearables. Nanostructured alloys, however, offer lower thermal conductivity and generally higher *zT* in Bi_2Te_3 alloys, allowing for the tuning of the peak *zT* to a desired temperature. In fact, the *zT* of Bi_2Te_3 has approached a saturation point, making further significant improvement difficult. While optimizing individual properties through new processing techniques like microwave processing [1, 51, 104] or dopant tuning [73] is still practical to slightly enhance Bi_2Te_3 performance for body heat harvesting, major power and voltage improvements ultimately require the discovery of new materials.

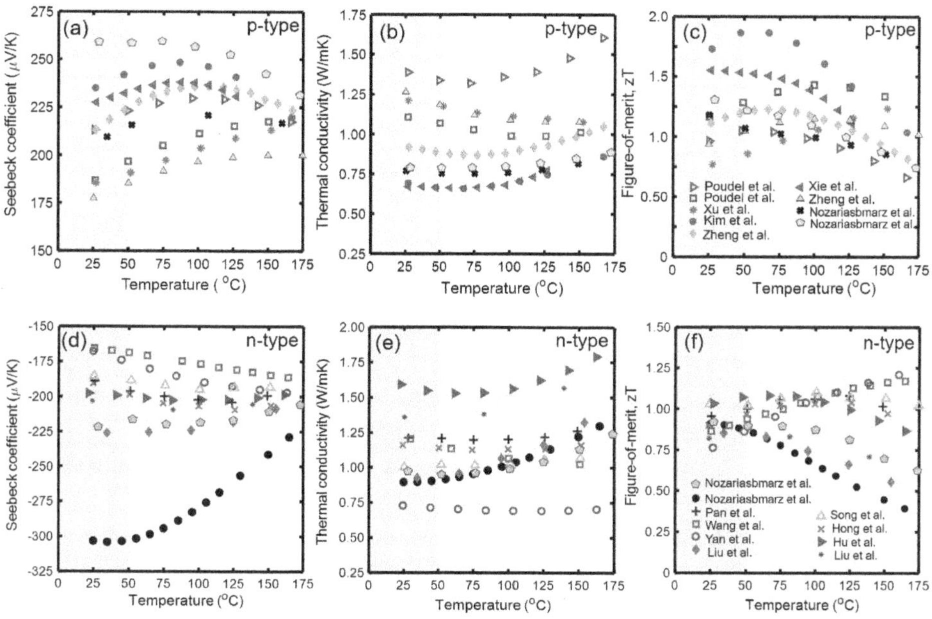

Figure 5.1: Comparison of Seebeck coefficient, thermal conductivity, and zT of the state-of-the-art (a–c) p-type $(Bi_xSb_{1-x})_2Te_3$ materials [51, 73, 90–95] and (d–f) n-type $Bi_2Te_{3-x}Se_x$ materials [1, 95–103] as a function of temperature in the range of 25 °C to 175 °C. The highlighted area in each graph demonstrates the typical operating temperature range for a wearable device.

Recently, Mg_3Bi_2-based [105] and Mg_3Sb_2-based [106] materials have demonstrated properties comparable to Bi_2Te_3 alloys, making them promising alternatives. However, they are known to suffer from poor thermal stability and cycling performance. Furthermore, there is a need for more research to specifically evaluate their effectiveness for body heat harvesting applications. Halide perovskites are also a new class of TE materials that have the potential to be used for wearable applications due to their superior high Seebeck coefficient and very low thermal conductivity [107].

5.2 System on chip for wearables

Wearable devices operate within a system that includes sensors (such as electrocardiogram and heart rate sensors), as well as data analysis and transmission components. These elements work together to collect, interpret, and wirelessly transfer data, with all of them requiring power to function efficiently within the wearable platform. Figure 5.2a illustrates the block diagram of a TEG-powered system-on-chip (SoC) platform. The system consists of several main components, including a power management unit (PMU) for power regulation; digital control, processing, and management modules that adjust system outputs using real-time feedback, analyze signals, implement algorithms, coordinate resources, and optimize overall system operation; an analog front-end (AFE) for converting weak, real-world analog signals from the sensors into a digital format; and a radio module for the wireless transmission of data.

5.2.1 DC–DC boost converters

One of the main challenges of using body energy harvesting is the fluctuations of energy, which can negatively impact device performance. For example, if a sensor like an ECG is directly connected to a TEG, a significant change in body heat flux could interrupt data acquisition. This is because a person's skin temperature constantly changes with their activity level, causing an unsteady heat flux. As a result, TEGs produce unstable and low power and voltage output, which is a critical consideration for the design of all wearable systems that rely on them.

TEGs produce a low-voltage direct current (DC) from body heat, typically ranging from a few to a few hundred millivolts. This voltage is not sufficient to power a wearable system directly. Therefore, this low voltage must be boosted by an electronic circuit called a DC–DC boost converter, which is a crucial component of the PMU. The DC–DC boost converter is made of a combination of a diode, an inductor, a switch, and a capacitor, and it is directly connected to the TEG to amplify its output voltage to a usable level [1, 108, 109].

5.2.2 Power management unit (PMU)

The PMU is a critical component in an integrated electronic system that acquires, controls, manages, and distributes electrical power within a device. Such a module is vital for low-powered devices, such as wearables, that rely on energy harvesting and battery-operated systems. The PMU primarily regulates the voltage from energy harvesting units, typically using a DC–DC boost converter, to achieve the stable voltage levels required by the device. It seamlessly connects various power sources, automatically switching among them to ensure a consistent power supply. The PMU also controls the charging and discharging of the battery from both energy harvesters and external power sources. Finally, the PMU optimizes energy consumption by turning off unused sensors and units, which helps to extend the device's operational time.

The efficiency of the PMU is directly related to the input voltage from the TEG, with efficiency increasing as voltage rises. For instance, efficiency is only 21% for 10 mV, but it jumps to 83% for 300 mV [1, 110, 111]. This clear relationship, illustrated in Figure 5.2b, shows that TEGs with a large output voltage (i.e., a high Seebeck coefficient) are required to maximize the PMU's efficiency in a wearable platform.

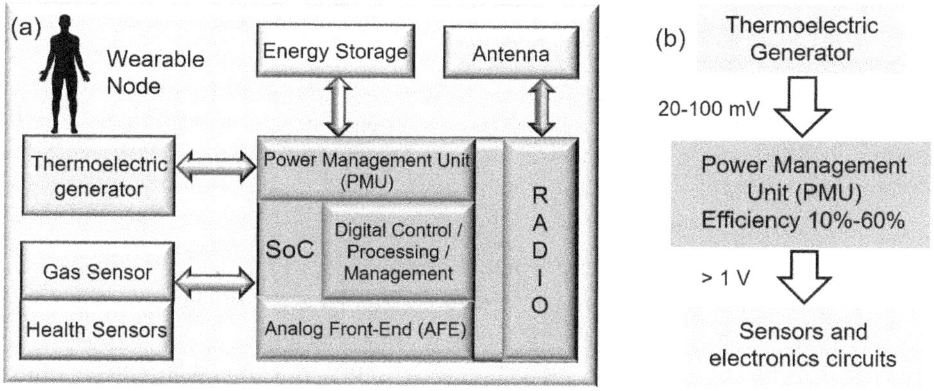

Figure 5.2: (a) Block diagram of a system-on-chip (SoC) platform consisting of a power management unit (PMU), digital control/processing/management modules, an analog front-end (AFE), a radio interface, a thermoelectric generator (TEG), sensors, and an energy storage component. (b) Simplified schematic illustrating the TEG output voltage and its interaction with the PMU. Conventional wearable TEGs typically produce low output voltages that cannot directly power sensors and electronic circuits. Although a PMU can boost the voltage, its efficiency is limited at low input levels. Consequently, there is a growing need for TEGs capable of generating higher output voltages. Adapted from Ref. [1], with permission from Elsevier.

The comprehensive model, which has been extensively discussed in Chapter 4, divides the TEG into infinitesimal sections to precisely calculate heat transfer within each region and the heat flow between them. This method provides a precise formulation for

TE device performance and design. The calculations for TEG output voltage, output power, and PMU power were performed for a TE leg with a base area of 0.25 mm^2, a leg length of 3 mm, and a fill fraction (FF) of 25%.

Figure 5.3a shows a 3D plot of TEG output power (P_{TEG}) as a function of thermal conductivity and Seebeck coefficient, specifically for an n-type material. While P_{TEG} increases with decreasing thermal conductivity, it shows little change with the Seebeck coefficient (Figure 5.3a and b). Therefore, a higher absolute value of the Seebeck coefficient does not significantly affect TEG output power.

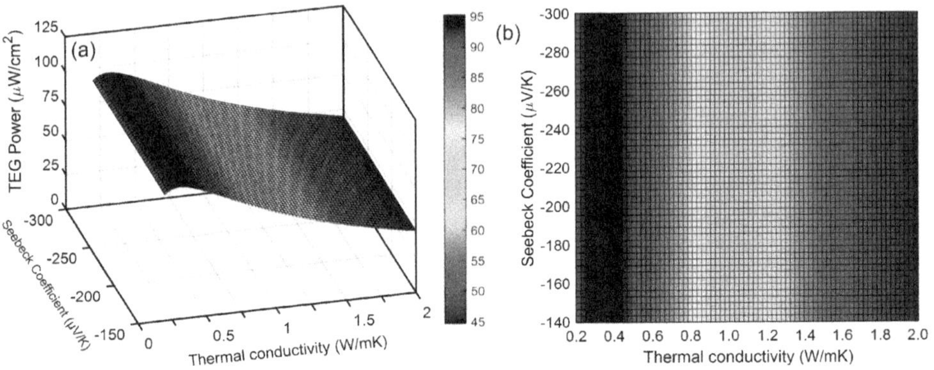

Figure 5.3: (a) A 3D plot of TEG power as a function of Seebeck coefficient and thermal conductivity (κ), with (b) its corresponding projected top view. All calculations were performed for a TEG with a leg cross-sectional area of 0.25 mm^2, a leg length of 3 mm, and a fill fraction of 25%. Reproduced from Ref. [1], with permission from Elsevier.

Figure 5.4a illustrates how the PMU output power (P_{PMU}) is affected by thermal conductivity (κ) and zT. Higher P_{PMU} is achieved with a higher zT and lower κ. For example, at a thermal conductivity of $\kappa = 0.4$ W/m · K, increasing zT from 1 to 2 boosts P_{PMU} by approximately 30% (from 47 μW/cm^2 to 61 μW/cm^2). In contrast, at a higher thermal conductivity of $\kappa = 2$ W/m · K, the same increase in zT only yields an 11% power increase (from 27 μW/cm^2 to 30 μW/cm^2). At a constant zT, while low thermal conductivities are generally desired for higher power, excessively low thermal conductivity (below 0.4 W/m · K) can block heat flow from the skin to the TEG, reducing P_{PMU}. As a result, the maximum P_{PMU} for a given thermal conductivity occurs at approximately $\kappa = 0.4$ W/m · K, as shown in (Figure 5.4b).

The P_{PMU} is also significantly influenced by the Seebeck coefficient. Surprisingly, a large region exists where a combination of small Seebeck coefficients and large thermal conductivities generates no power. In Figure 5.4c, the dark blue area represents this critical configuration where P_{PMU} is zero. There is a linear relationship between the critical Seebeck coefficient and thermal conductivity. Figure 5.4d further demonstrates that TE materials with lower thermal conductivity and higher Seebeck coefficients produce a greater P_{PMU}. At higher Seebeck coefficients, the power output of the wearable system becomes less sensitive to variations in thermal conductivity. Therefore, both the Seebeck coefficient and thermal conductivity are key material parameters that must be carefully optimized to achieve optimal TEG performance in a wearable system.

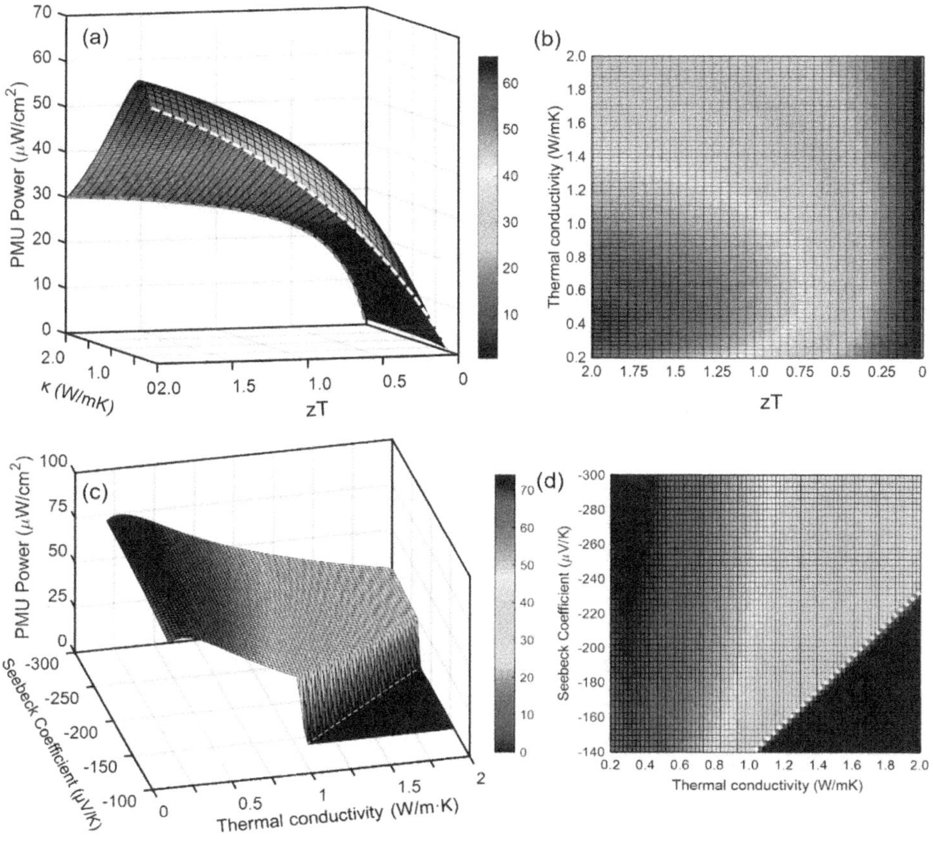

Figure 5.4: (a) A 3D plot of the power management unit (PMU) power as a function of thermal conductivity (κ) and zT. (b) A projected top view of (a) showing that the maximum power achieves at lower thermal conductivities. (c) A 3D plot of PMU power as a function of the Seebeck coefficient and κ. (d) A projected top view of (c) identifying the combination of materials properties that produce no power. All calculations were done for a TEG with a TE leg cross sectional area of 0.25 mm^2, a leg length of 3 mm, and a fill fraction of 25%. Reproduced from Ref. [1], with permission from Elsevier.

5.3 Wearable energy storage

A self-powered wearable device is expected to continuously generate power from the human body. However, the amount of power produced can fluctuate depending on physical activity, timing, and the specific energy harvesting method used. At certain moments, the generated power might not be enough to run the electronics. Therefore, for reliable operation, a portion of this power needs to be stored. For example, when a person is inactive and device power requirements are minimal, the extra energy can be saved for later use. Beyond storage, power regulation is critical for optimal device performance. Instead of directly connecting a sensor to a TEG, a PMU is used to provide, manage, and control power for various components. Without energy storage, the device would only function when energy is actively being harvested and available. Ideally, power should be saved in a separate storage unit, and the PMU should draw from that source as needed to provide smooth, regulated voltage and current to the device. Figure 5.5 shows the power requirements for various wearable sensors and implantable devices [112].

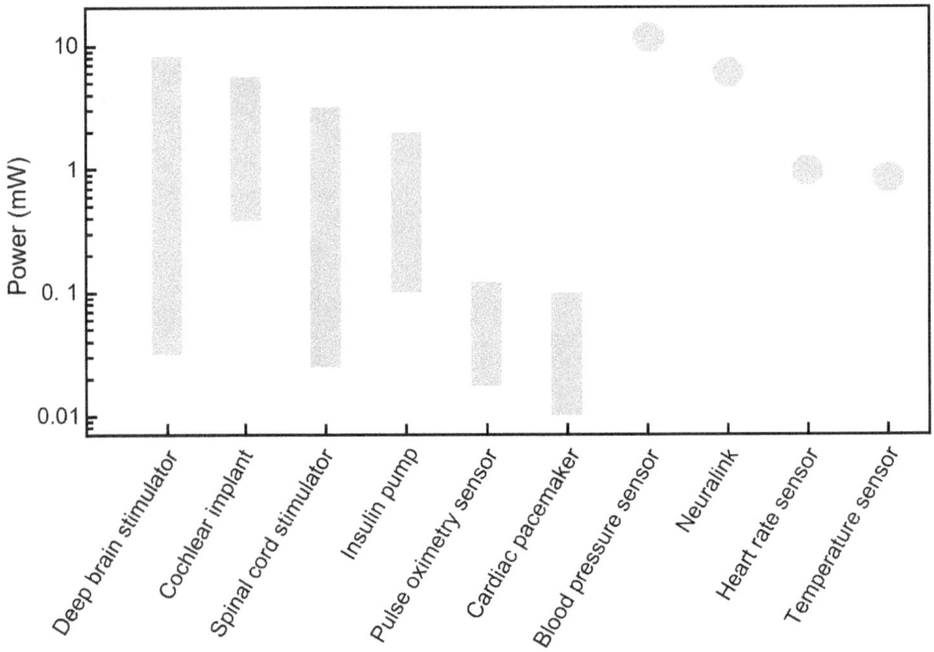

Figure 5.5: Power requirements for various wearable sensors and implantable devices, including deep brain stimulators, cochlear implants, spinal cord stimulators, insulin pumps, pulse oximetry sensors, cardiac pacemakers, blood pressure sensors, Neuralink, heart rate sensors, and temperature sensors. Adapted from [112].

Energy can be stored using two different methods: rechargeable batteries and supercapacitors. Rechargeable batteries use reversible electrochemical reactions, specifically oxidation and reduction, to generate electricity. This allows them to be charged, discharged, and recharged multiple times. During discharge, chemical energy is converted into electrical energy at the battery's electrodes. When charging, electrical energy is used to restore the chemical energy state. A rechargeable battery typically consists of a negative electrode (the electron donor), a positive electrode (the electron acceptor), an electrolyte for ion movement between the electrodes, and a separator to keep the electrodes apart from each other (Figure 5.6a). The most common types of rechargeable batteries are lithium-polymer (Li-Po), lithium-ion (Li-ion), solid-state lithium, and zinc-based batteries.

Supercapacitors are rechargeable energy storage devices that store energy through electrostatic charge separation, giving them high capacitance and a fast charge and discharge rate. They have higher capacitance than conventional capacitors and can be recharged faster than batteries. Like batteries, supercapacitors have positive and negative electrodes, an electrolyte, and a separator (Figure 5.6b). However, their materials and functionality are different. Instead of relying on chemical reactions, supercapacitors store energy by physically accumulating opposite charges on the surface of their electrodes.

In general, compared to supercapacitors, Li-ion batteries have a higher energy density, lower power density (Figure 5.7), longer charge time, lower life cycle, lower cost, and a faster degradation rate over time. For wearable applications, batteries are a good choice when a lot of energy is needed for extended periods. However, supercapacitors are ideal for applications that require fast charging and discharging, and a long lifespan [113].

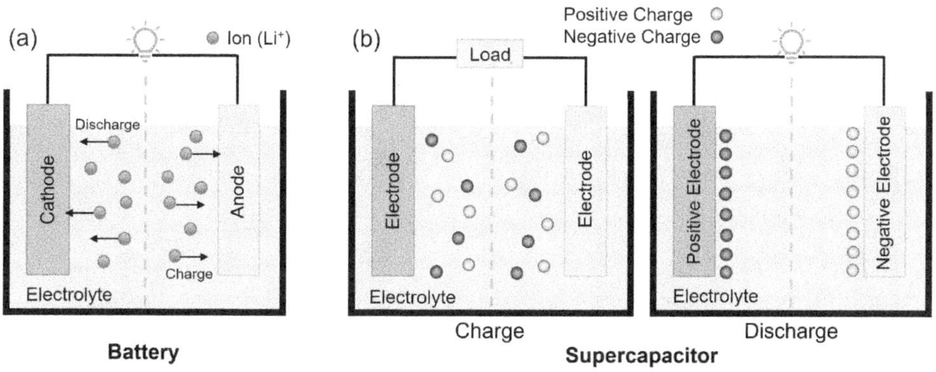

Figure 5.6: Schematic drawings of wearable energy storage techniques: (a) batteries and (b) supercapacitors.

There is growing interest in developing lightweight and flexible energy storage solutions to power the next generation of wearable devices. The rigid and bulky nature of current commercial batteries and supercapacitors makes them difficult to integrate into miniaturized wearables. This limitation restricts both the devices' wearability and their ability to conform to the body's surface. For example, fiber-based or wire-shaped energy storage devices are a promising alternative because they can be woven directly into electronic textiles to meet the practical needs of wearable technology. The ideal energy storage technology for wearables must meet several key requirements, including long-term durability under mechanical stress, high power density, wearability, comfort, washability, stretchability, and scalability.

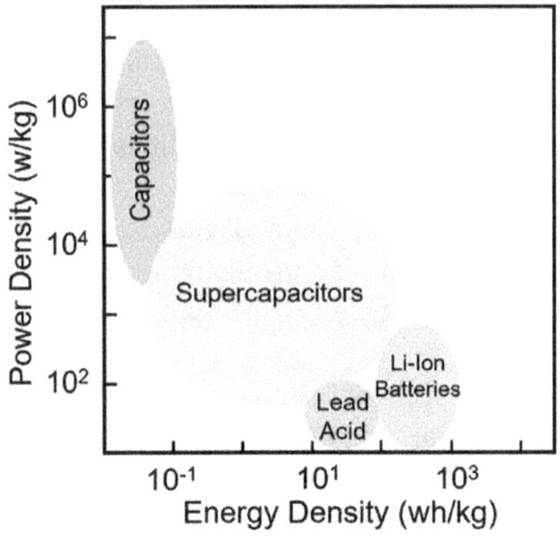

Figure 5.7: Comparison of power and energy density of different energy storage techniques. Adapted from [114].

6 Rigid thermoelectric device fabrication methods

As discussed in Chapter 3, high electrical contact resistance at the junction of a thermoelectric (TE) leg and metal interconnects limits the performance of TE devices, specifically when they are used as thin films. Thicker TE materials can compensate for this issue to some extent. Although recent research has shown improved performance in TE thin-film materials and devices [115, 116], with current technology, most TE materials are made through bulk processing. Bulk TE materials have demonstrated reliable, repeatable, and improved performance. Wearable thermoelectric generators (TEGs) are generally divided into two main categories: rigid and flexible devices. Rigid TEGs are made from bulk TE materials, while flexible TEGs are either made of bulk materials on a flexible substrate or are comprised of flexible materials and devices. This chapter discusses the fabrication methods for rigid TE materials and devices used in wearables.

6.1 Fabrication processing of bulk thermoelectric materials

The quality of TE materials depends on materials synthesis and processing techniques, the purity of the raw materials, and the cleanness of the process. Additionally, the simplicity, repeatability, and lower cost of the process are critical. In general, any materials synthesis technique used for metals, alloys, semiconductors, ceramics, and polymers can be applied to TE materials. Most TE materials are considered as electronic materials, and they are highly sensitive to purity, dopant, and oxidation during the fabrication process. Synthesis of bulk TE materials is divided into two main categories: single-crystal growth and powder processing/melting techniques. Each of these categories has several fabrication processing techniques to develop bulk TE materials. The most important and reliable techniques will be discussed below.

6.1.1 Single-crystal growth

One of the primary techniques for fabricating TE materials is through single-crystal growth. There are several common techniques for growing single crystals of TE materials: (1) Bridgman technique: In this method, the material is melted inside a crucible, such as quartz, and is then slowly solidified by moving one side through a temperature gradient. This method can be used to fabricate single crystals of PbTe, Bi_2Te_3, Mg_3Bi_2, SnSe, and GeTe (Figure 6.1). High TE properties have been achieved in Mg_3Bi_2 [117] and SnSe [118] alloys. (2) Czochralski technique: The process begins by inserting a seed crystal into a molten pool of the material and slowly lifting it upward. A single crystal then begins to grow on the seed. While this is the main technique for commer-

 | https://doi.org/10.1515/9781501514814-006

cially fabricating high-quality silicon single crystals, it can also be used to grow single crystals of SiGe and Bi_2Te_3.

Other less common single-crystal growth techniques for TE materials include floating zone and flux growth. Single-crystal TE materials have minimal scattering of phonons and charge carriers because they lack grain boundaries, which are a major source of scattering. As a result, these materials have the highest carrier mobility and a minimal defect density.

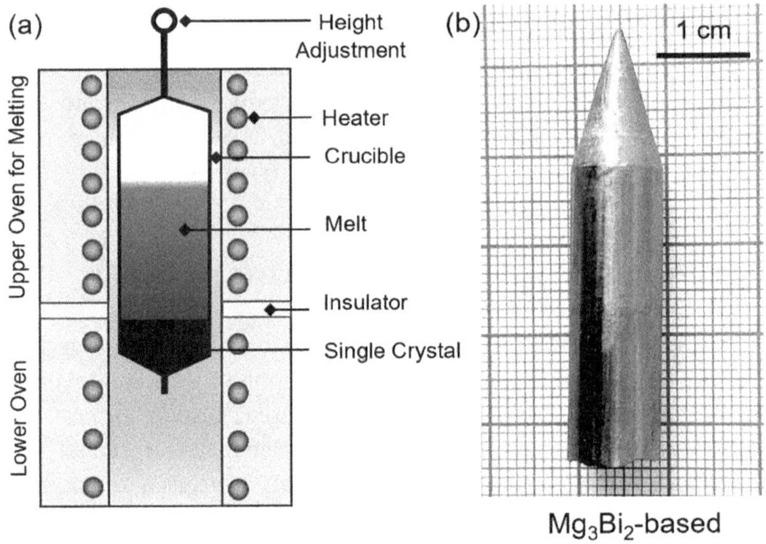

Figure 6.1: (a) A schematic drawing of Bridgman technique for single-crystal growth. (b) An actual picture of a grown single crystal of $Mg_3Bi_{1.49}Sb_{0.5}Te_{0.01}$. Reproduced from Ref. [117], with permission from Elsevier.

6.1.2 Powder processing and melting

Powder processing, melting, and combinations of these methods are the most common techniques for discovering, fabricating, and processing TE materials. Compared to single-crystal growth, which involves the complex fabrication of a uniform, defect-free single crystal, powder and melting techniques can synthesize various compounds in a relatively fast and easy manner. The primary techniques used to synthesize and consolidate TE materials will be discussed briefly below.

6.1.2.1 Mechanical milling/alloying

Mechanical milling/alloying, also known as ball milling, is a solid-state technique used to synthesize most TE materials. In this technique, a stoichiometric mixture of raw elements is weighed and loaded into a vial (typically made of hardened steel, zirconia, or tungsten carbide) along with balls, at a known ball-to-powder ratio. The vial is sealed inside an inert-gas-filled glovebox to ensure a controlled environment and prevent oxidation. The loads are subsequently milled for a specific amount of time, from a few minutes to several hours, using ball-milling equipment. The mechanical force and impact from high-speed balls that collides with the moving jar (with rotating, vibrating, or planetary motion) can grind, mill, and mix the mixture of raw materials and/or induce chemical reactions to synthesize an alloy or compound. The collisions between the balls and the jar apply friction, impact, and shear force to the material. In addition to creating the compound, this technique can also reduce the particle and grain sizes of the materials, and it is one of the most widely used methods for fabricating nanostructured materials in various fields. Figure 6.2 a and b shows the working principles of ball milling. In general, this technique is mainly applied to synthesize

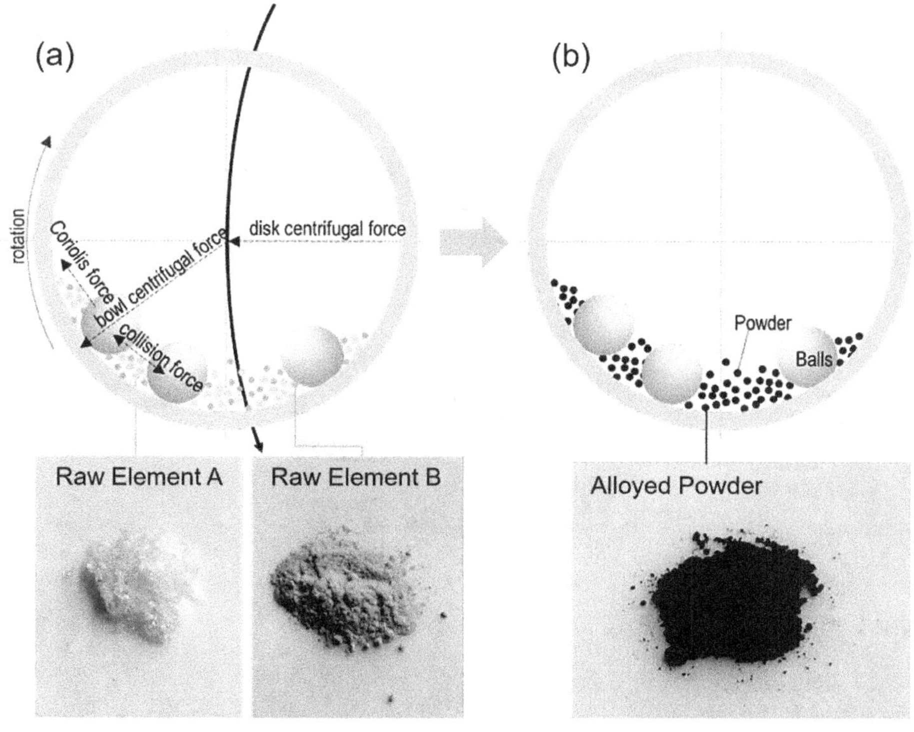

Figure 6.2: Working principles of mechanical milling/alloying. (a) Multiple forces are applied to the mixture of powders. (b) The resulting conversion of raw elements into a uniform alloyed powder. Adapted from [107] licensed under CC BY 4.0.

new compounds (via mechanical alloying) or reduce particle (or grain) sizes (via mechanical milling). For wearable applications, high-performance bismuth telluride alloys and compounds have been successfully synthesized using ball milling technique.

6.1.2.2 Melting techniques

Another widely used method for fabricating TE materials is based on melting techniques. The raw elements are weighed at their stoichiometric ratio and poured into a crucible made of graphite, tungsten, tungsten carbide, zirconia, alumina, or boron nitride. Through various techniques, the material undergoes melting with subsequent quenching. Melting can be performed using an induction furnace, arc melting, or a resistance melting furnace. In an induction furnace, an electromagnetic field generates eddy currents in a conductive material, generating heat through resistive heating to melt the mixture and create a uniform compound (Figure 6.3). In arc melting, an electric arc between a tungsten electrode and the material melts the raw elements, resulting in a uniform compound. In resistance melting, an electrical current heats the heating elements of the furnace to achieve a high temperature in a closed environment. The raw material is heated through conduction, convection, and radiation until it reaches the melting temperature of the mixture of compounds. The melted ingot can be used as is or crushed and pulverized into a powder for consolidation into a specific shape and dimensions.

Figure 6.3: Induction melting process and the melting ingot.

6.1.2.3 Consolidation

The powders from mentioned techniques can be sintered into a pellet using several methods, including green body sintering, hot pressing, and spark plasma sintering (SPS). Green body sintering is the simplest technique. Here, powder is cold pressed in a die, and the resulting pellet is then placed in a furnace at high temperature to sinter.

However, because no pressure is applied during sintering, the material cannot achieve its full density, typically reaching only 70% to 90%. In hot pressing, powder is placed in a die surrounded by a heater or furnace. Pressure is continuously applied to the powder by two plungers while the die heats up. This method allows the material to achieve density close to 100%.

SPS, also referred to as field-assisted sintering technique (FAST), is the most common method for consolidating TE materials. In SPS, powder is inserted into a conductive graphite die where two plungers apply pressure from the top and bottom. A pulsed DC current is passed directly through the die, heating the powder via Joule heating. At higher temperatures, as the compacted powder softens, a uniaxial press maintains pressure on the sample. The heating and cooling processes is controllable and can be extremely fast or slow, and the sample can be held at a specific temperature to achieve optimal results. SPS offers precise control over sintering temperature, time, pressure, and atmosphere. Figure 6.4 a-c shows the SPS setup, the process of sintering, and an actual picture of a graphite die, which displays the black body radiation at high temperatures during SPS.

6.1.2.4 Other techniques

Melt spinning: In this method, an alloy is melted in a furnace (typically via induction melting) and then ejected from a nozzle onto a rapidly rotating, water-cooled wheel. This process forms thin ribbons with a specific aspect ratio. The extremely fast cooling rates result in amorphous structures, and the ribbons must be further consolidated to make a bulk sample.

Hot extrusion: This method involves subjecting a billet, made of pressed powder or pre-sintered material, to plastic deformation through a die with a smaller cross-sectional area to create a uniform, long rod. The billet is heated to about 50% to 75% of its melting temperature to increase its plasticity. Extrusion results in a highly dense material with improved mechanical properties and grains that are aligned in the direction of extrusion. This technique is ideal for BiTe and PbTe TE alloys, as it produces a uniform, mechanically enhanced rod for further device fabrication.

Self-propagating high-temperature synthesis (SHS): SHS is a simple and effective combustion technique used to synthesize various materials. In this method, an initial mixture of a compound is weighed and mixed according to its stoichiometric ratio and then compacted into a green body. An exothermic reaction is then ignited using a localized heat source, such as a spark or hot wire. Once ignited, a reactive wave sustainably propagates through the material. The temperature of this reaction can reach up to 3,500 °C in a matter of seconds. The products of SHS are typically porous and require a secondary consolidation process to achieve a fully dense material.

Microwave processing: This is a newer technique that uses microwave as an electromagnetic wave in either a single-mode (Figure 6.5) or multi-mode cavity to heat up TE materials. The initial material can be a stoichiometric mixture of a compound or a

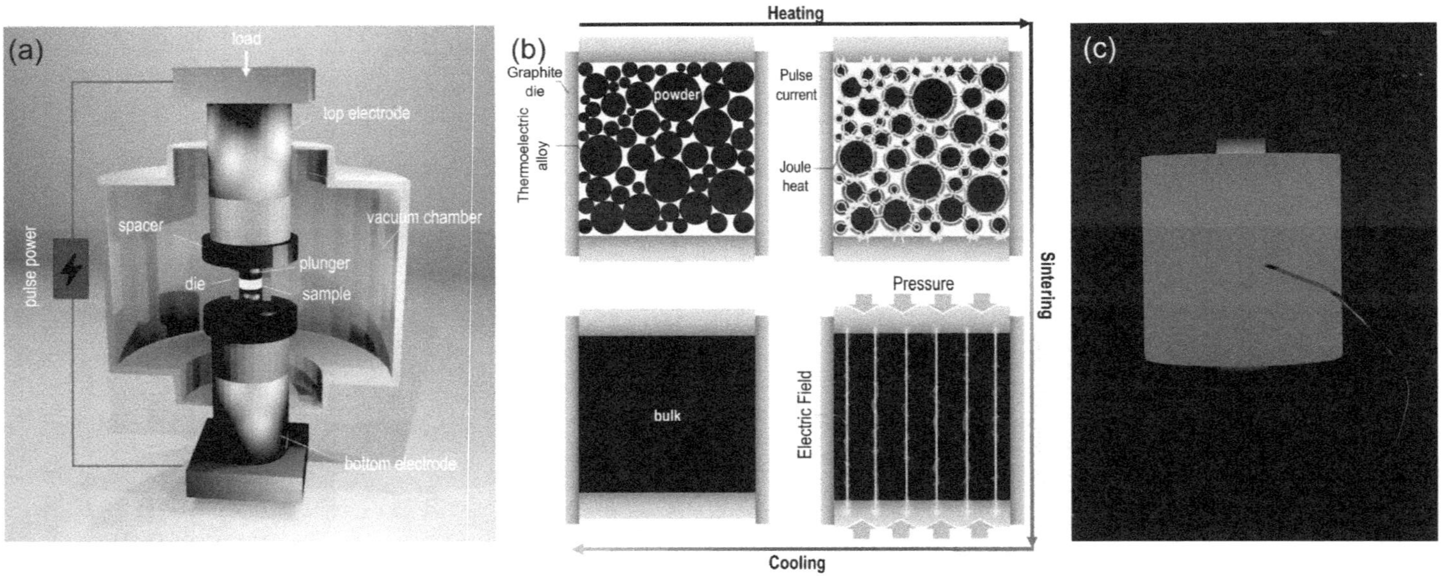

Figure 6.4: (a) Schematic setup of spark plasma sintering (SPS) showing thermoelectric powder being pressed inside a graphite die with two plungers. (b) Processing diagram of the sintering process, illustrating how powder particles are sintered. (c) Actual picture of SPS graphite die, showing the black body radiation at high temperature ((a) and (b) reproduced from [107] licensed under CC BY 4.0).

uniform compound in powder, pre-sintered, or fully sintered form. Microwave processing enables in situ synthesis, sintering, and consolidation. The nonthermal effects of microwave facilitate the discovery of new TE compounds and the optimization of their properties. Compared to other methods, microwave processing for the synthesis of TE materials is a newer technique, but it shows great promise for in situ processing.

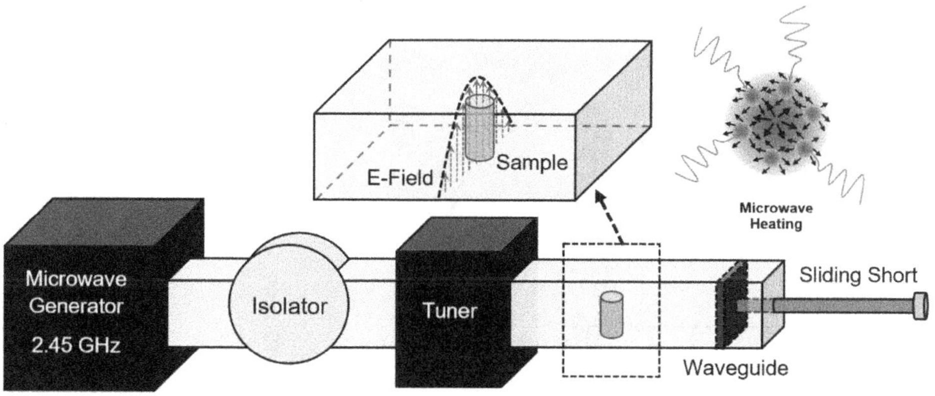

Figure 6.5: Schematic diagram of a single-mode cavity microwave processing setup. The bulk sample is positioned at the center of the waveguide and irradiated with high-energy microwave electric field. Unlike conventional heating where the surface of a material is heated and heat is conducted inward, microwave processing heats all parts of the material uniformly, as shown in the top right drawing. Reproduced from Ref. [1], with permission from Elsevier.

After synthesis and consolidation using discussed techniques, all pellets must be polished or sliced to the desired height. Both sides of the pellets are then metallized using a sophisticated technique to create a reliable metallic surface for subsequent contact and soldering. The metallized pellets are then diced into TE legs. For rigid wearable TEGs, the leg height can range from a few hundred micrometers to several millimeters. The details of bulk TEG devices and their applications are discussed in Chapter 3.

6.2 Rigid thermoelectric generators

Rigid TEGs are solid, non-bendable, and non-deformable TEGs that are widely available commercially. In these devices, all headers, legs, and junctions are solid. The power density of rigid TEGs ranges from hundreds of nW/cm^2 to a few tens of μW/cm^2 under no external cooling. The on-body performance of rigid TEGs is significantly affected by the heat sink material, shape, size and design. Additionally, the use of heat spreaders to collect heat from the hot side and dissipate it from the cold side is a cru-

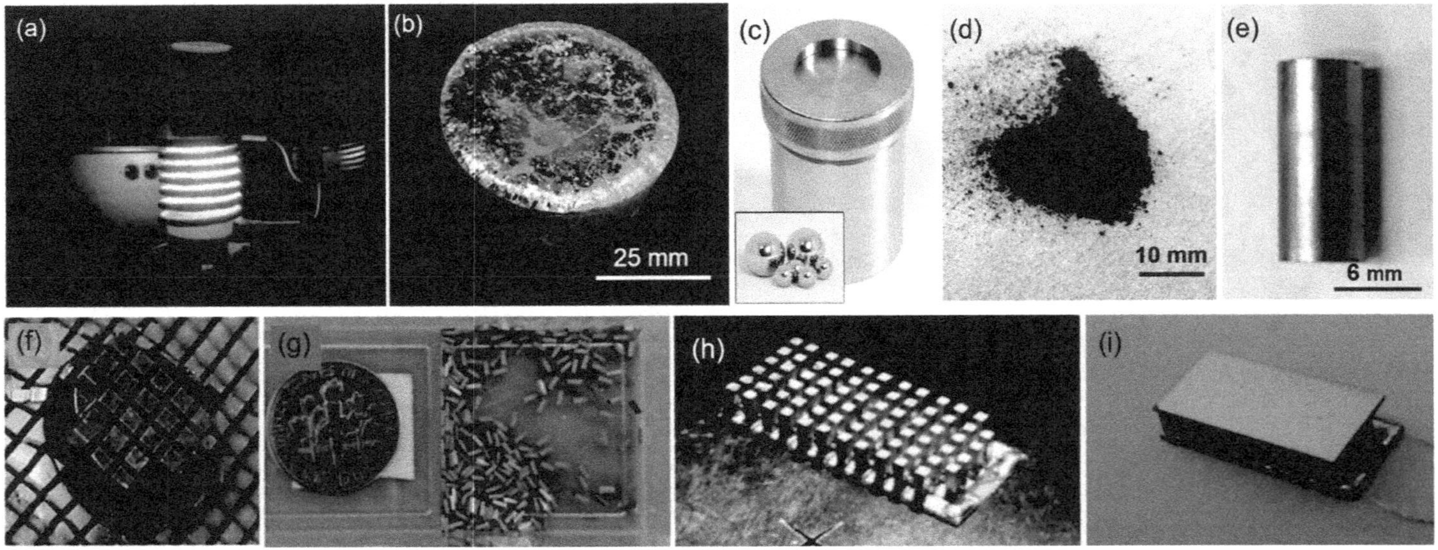

Figure 6.6: Raw elements to wearable device fabrication flow: (a) induction melting of the raw elements, (b) a BiTe alloy ingot produced from induction melting, (c) a ball milling jar and balls, (d) BiTe nano-powders after ball milling, (e) consolidated BiTe using spark plasma sintering (SPS) [119], (f and g) diced thermoelectric legs, (h) packaging the legs into a device, and (i) a TEG device. (f–i) Reproduced from Ref. [51], with permission from Elsevier.

cial factor. Figure 6.6 shows the fabrication steps for a wearable TEG from raw elements to the final device. The fabrication process for rigid TEGs may include fewer or additional steps.

Without a heat sink or heat spreader, commercial TE materials and devices typically generate a power density less than 10 μW/cm^2. Leonov has performed extensive research on optimizing TEG designs for wearable applications and reported a power density of 20 μW/cm^2 from wrists using Bi_2Te_3 TEGs and a large heat sink [120–122]. Nozariasbmarz et al. have done thorough research on design, synthesis, and fabrication of p- and n-type BiTe-based alloys and devices specifically optimized for wearable applications [1, 51, 70, 73, 75, 76, 83, 107]. Two key factors in TEG design are fill fraction (FF) and aspect ratio (AR). Commercial off-the-shelf (COTS) TE modules are typically fabricated with an FF in the range of 25–60% and an AR of 1–2. While rigid TEGs with an FF below 15% may suffer from poor mechanical stability, an optimized design can uniformly distribute the mechanical load and improve durability. Experimental results on the effect of FF show that an optimized module with FF of 12% can generate 34 μW/cm^2, which is up to 80% more power than a module with FF of 36% under steady-state conditions (Figure 6.7a). This FF optimization also decreased the cost and weight of the module. However, there is a trade-off that a lower FF design yields higher output power, but a higher FF design produces a higher output voltage. As a higher number of TE legs results in higher output voltage, and as mentioned in Chapter 5, a higher output voltage is required for better PMU power efficiency.

Upon being placed on the skin, the temperature profile of the hot and cold sides of the rigid TEG shows that the local skin temperature initially drops below 30 °C before stabilizing at a higher temperature. This phenomenon creates an initial peak in the generated power and voltage from the TEG, which then saturate at a lower values (Figure 6.7b). The hot and cold side temperatures, as well as the temperature difference (ΔT) across the TEG, can vary depending on the skin temperature of different body parts and the specific design of the TEG. Similarly, the heat sink temperature profile changes, as it is attached to the body. Initially at ambient temperature, the heat sink temperature rises until it reaches an equilibrium point where the heat being rejected to the environment equals the heat being conducted from the body to the heat sink.

Airflow is another crucial factor for TEG performance because it enhances heat rejection from the cold side (typically a heat sink) to the ambient air, resulting in a higher ΔT across the TEG. For example, a COTS rigid TEG produces a power density of just 5.7 μW/cm^2 on a wearable platform without a heat sink or airflow. However, using forced air convection increases the power density of the same COTS TEG to 35.5 μW/cm^2 [51].

To maximize the performance of a wearable TEG, the TE properties of its materials must be optimized. As mentioned previously, maximum performance is theoreti-

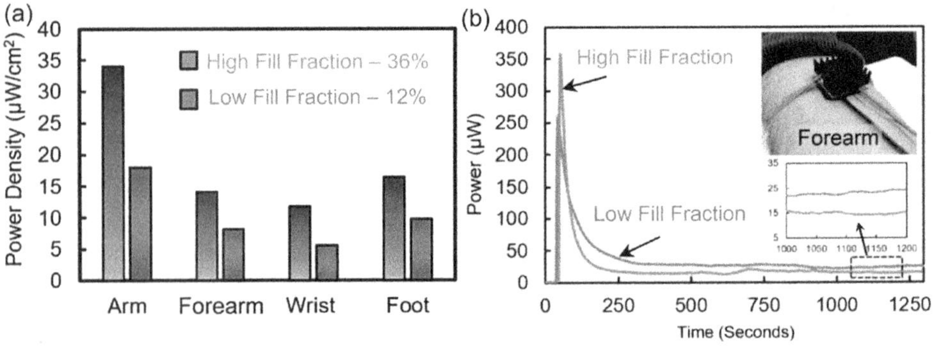

Figure 6.7: (a) Comparison of the power density generated from the arm, forearm, wrist, and foot for TEGs with low (12%) and high (36%) fill fractions. (b) Transient output power versus time for low and high fill fraction TEGs, showing the initial peak of power immediately after the TEG is applied to the skin. The inset displays an actual picture of the setup on the forearm. Reproduced with permission from Ref. [70]. Copyright © [2019] American Chemical Society.

cally expected from materials with a high zT and low thermal conductivity. For instance, nanocomposite materials with low thermal conductivity (~0.77 W/mK) and a zT of ~ 1.2 perform significantly better than commercial materials, which generally have high thermal conductivity. TEGs made from a nanocomposite BiTe alloy generated up to 44.2 μW/cm^2 and 49.7 mV/cm^2 without a heat sink or airflow. When tested on the body with increased airflow, the power density for the nanocomposite material reached a maximum of ~156.5 μW/cm^2. This is 4 to 7 times higher than COTS TEGs of a similar size under the same conditions (Figure 6.8a).

The performance of a TEG in a wearable armband was demonstrated by connecting the TEG to a DC-DC boost converter and an LED lamp (Figure 6.8b). The LED turned on as soon as the wearable TEG was placed on the skin (Figure 6.8c). Table 6.1 summarizes the key results from early and significant studies on body heat harvesting using rigid TEGs.

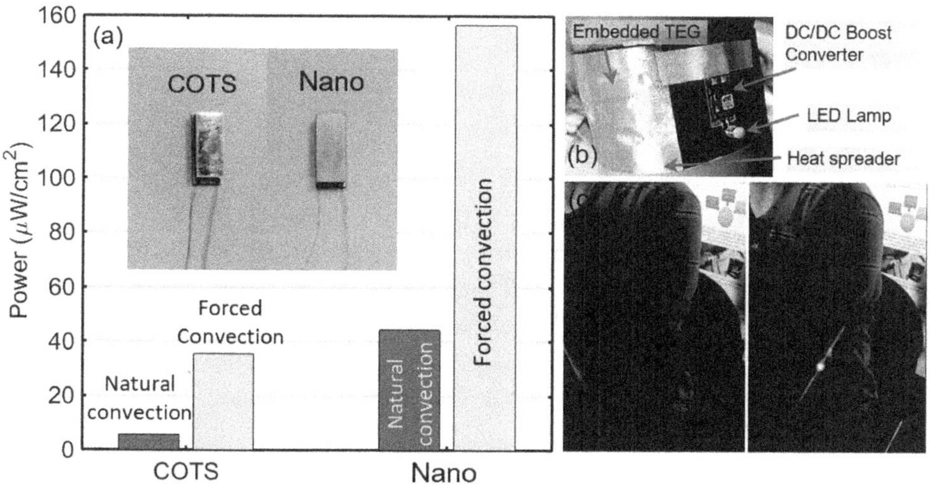

Figure 6.8: (a) Comparison of the generated power from a COTS TEG and a device made of nanostructured (nano) materials. The inset shows actual pictures of the two devices. The nano-TEG produced 300% to 680% more power than the COTS device of similar size under the same conditions on the body. (b) A wearable energy harvesting armband device with nano-TEGs, an LED lamp, a DC–DC boost converter, and a graphite heat spreader. (c) LED lamp turned on when the device was worn on body. Reproduced from Ref. [51], with permission from Elsevier.

Table 6.1: A summary of some of the pioneering and high-performance rigid TEGs for body heat harvesting reported in the literature.

	V_{oc} (mV/cm^2)	Power per cm^2	Year	Note	References
Strasser et al.	–	<100 nW	2004	No heat sink	[123]
Settaluri et al.	10.8	21.6 μW	2012	Thin Cu heat sink, no airflow	[124]
Leonov	–	20 μW	2013	Very large heat sink	[125]
Nozariasbmarz et al.	–	34 μW	2019	Large heat sink, no airflow	[70]

7 Flexible thermoelectric device fabrication methods

Flexibility of a device can enhance its wearability and comfort, reduce its weight, and improve mechanical properties such as twisting, bending, and stretching. While thermoelectric (TE) materials perform best in bulk form, flexible TEGs offer unique properties that make them ideal for future wearable applications. Flexible TEGs can be made from rigid materials on a flexible platform or be composed of fully flexible materials and devices. This flexibility can originate from the material itself, the interconnects, the headers, or a combination of these elements. This chapter discusses various techniques for fabricating flexible TEGs.

7.1 Flexibility versus stretchability

The field of flexible wearable devices involves both flexibility and stretchability, which are distinct mechanical properties. A material that is flexible is not necessarily stretchable. Flexibility is the ability of a device to bend or fold without breaking. In general, a thinner material is more flexible than a thicker one. According to solid mechanics, the bending stiffness (D) of a plate is given as follows [126]:

$$D = \frac{(Eh^3)}{12(1 - v^2)} \tag{7.1}$$

where E is Young's modulus of the materials, h is the thickness of the plate, and v is Poisson's ratio. Accordingly, bending stiffness is proportional to the cube of thickness. Thus, reducing the thickness of the material into half results in one-eighth the stiffness, or eight times more flexibility. That is how a metal foil becomes flexible. Stretchability is the ability of a device to stretch and be functional under tensile strain. Similarly, compressibility is the ability of a device to compress under compressive strain.

7.2 Device architectures

Figure 7.1 demonstrates various common architectures of flexible TEGs that have been widely used in the literature, including vertical, horizontal, hybrid, circular, Y-type, and monopolar structures [127]. The most widely used architectures are vertical and horizontal. In a vertical design, the p- and n-type TE legs are positioned perpendicular to the substrate, similar to conventional rigid TEGs. Conversely, in horizontal design, the legs are parallel to the substrate. The hybrid design combines both vertical and horizontal approaches. The circular design arranges the TE legs in a cylindrical pat-

 | https://doi.org/10.1515/9781501514814-007

tern, making it ideal for use on pipelines. The *Y*-type design uses *Y*-shaped TE legs combined with a metal heat conductor. The monopole design features vertically arranged TE legs of the same type (either *n*- or *p*-type) connected by metal components.

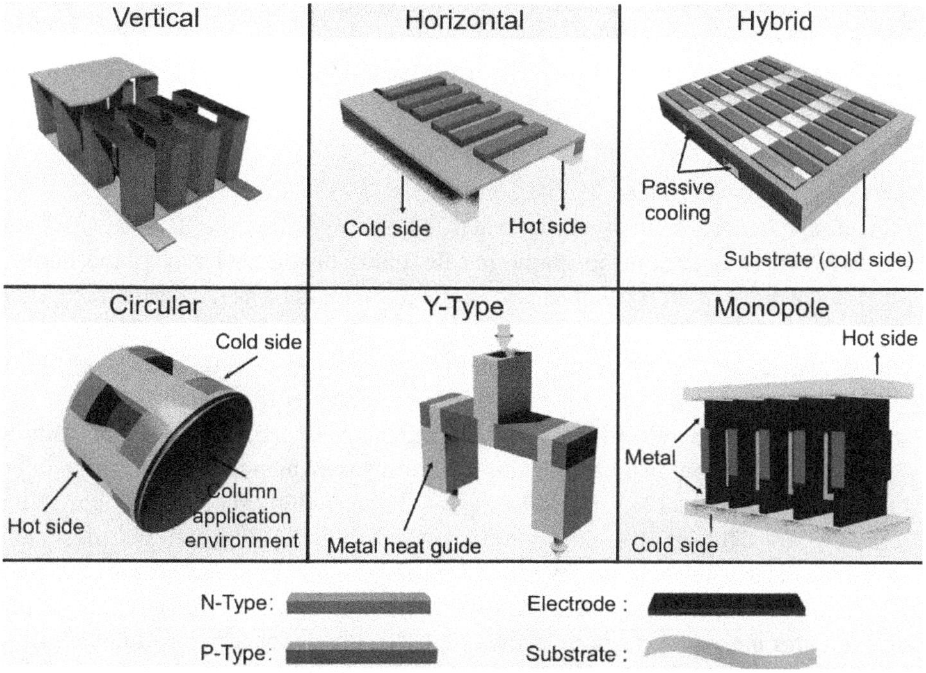

Figure 7.1: Different architecture of flexible TEGs including vertical, horizontal, hybrid, circular, Y-type, and monopole structures. Reproduced from Ref. [127], with permission from Elsevier.

7.3 Flexible TEGs from bulk materials

To date, bulk TE materials show much better properties than flexible or thin films. The simplest way to fabricate flexible TEGs is by embedding bulk TE materials, also known as rigid materials, in a flexible matrix using flexible interconnects. In TE devices, the legs are connected electrically in series and thermally in parallel. Therefore, the material filling the area between the bulk TE legs must be both an electrical and thermal insulator to maintain the TEG high performance. This method is ideal for creating a high-performance flexible TEG that combines the advantages of flexibility with the superior properties of bulk materials.

However, since these flexible devices are made from bulk materials, they face similar limitations. For instance, longer legs can improve the performance of wearable TEGs but reduce the flexibility of the device. A lack of flexibility and stretchability in the interconnects, as well as fatigue from repeated bending, can limit device performance.

7.3.1 Fabrication of flexible TEGs from bulk materials

Depending on the flexibility of the headers, interconnects, or the matrix holding the TE legs, different designs have been suggested:

7.3.1.1 Rigid TEGs embedded in a flexible matrix

In this design, a separately fabricated rigid TEG is embedded in a flexible substrate, such as PDMS, to maintain the shape and flexibility of the device. Top and bottom heat spreaders are in direct contact with the TEG to enhance heat collection from the body and heat dissipation to the surroundings. The spreader can be any flexible material with high thermal conductivity, such as carbon, copper, or aluminum foils.

Figure 7.2a shows a schematic of such device. It can be used on different parts of the body, including the wrist (Figure 7.2b), arm, and chest, or integrated into clothing to generate power from body heat. Figure 7.2c compares the power density generated from different body locations. Devices integrated into clothing generate less power because they lack tight contact with the skin. When a person is idle and there is no airflow on the device, power generation across different body locations is approximately 5 $\mu W/cm^2$. However, when a person is walking (i.e., at higher air velocity), the upper arm generates more power, up to 21 $\mu W/cm^2$ [128]. The upper arm location yields the highest power in this specific device design.

The power requirements for low-powered health monitoring sensors, such as an accelerometer, ECG, and ozone sensor, are 10 μW, 50 μW, and 150 μW, respectively. Therefore, properly designed rigid TEGs on a flexible platform can produce enough power to run these sensors. Although integrating the device into T-shirts or clothing reduces the output power, this design is the most comfortable for wearable applications while generating 2–8 $\mu W/cm^2$ of power. Additionally, it can be embedded into a larger surface area of clothing, which enhances the overall output power.

7.3.1.2 Rigid TE legs and interconnects with flexible headers

Another design uses bulk TE legs and interconnects with flexible headers. In this design, the TE legs are soldered directly onto a metal contact that has been deposited on a flexible substrate such as PDMS (Figure 7.3a). This design ensures ideal direct contact between TE legs, headers, and skin. However, the mechanical robustness at the interconnect junction may degrade after just a few bending cycles.

Low-temperature solders are ideal for this type of device because they are more compatible with PDMS. Commercial TE legs of any dimensions can be used to fabricate this device. Figure 7.3b shows the output power density versus airflow for such

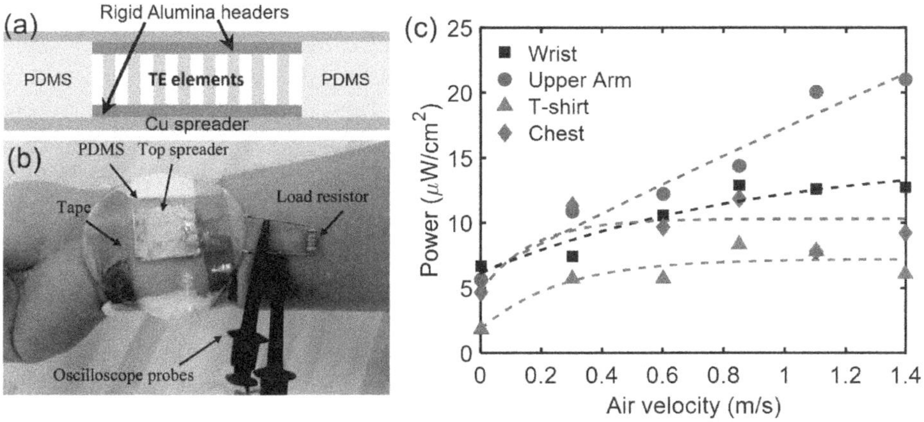

Figure 7.2: (a) Schematic design of a rigid TEG device embedded in a flexible PDMS substrate with a copper spreader. (b) Actual picture of the rigid TEG worn on the wrist. (c) Comparison of the power density on wrist, upper arm, T-shirt, and chest. Reproduced from Ref. [128], with permission from Elsevier.

devices. This device can generate up to 8 μW/cm^2 power at various airflow speeds ranging from 0 to 6 m/s [129,130].

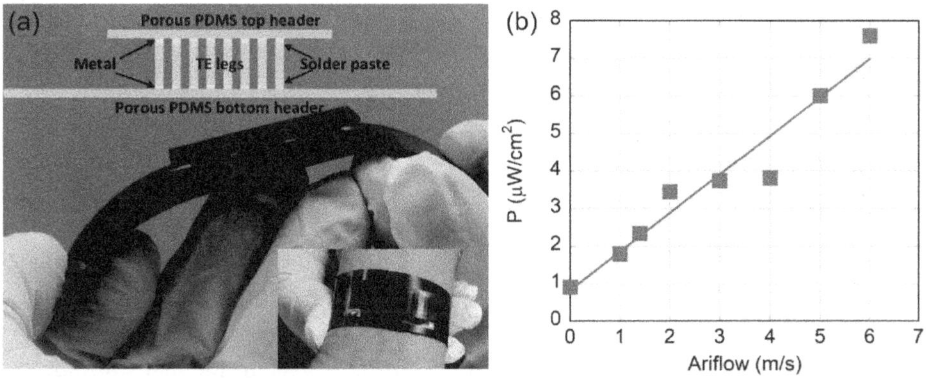

Figure 7.3: (a) A flexible TEG with rigid legs and interconnects, and flexible top and bottom headers made of PDMS. (b) The power density of the flexible TEG at various airflow speeds. Reproduced from Ref. [129], with permission from Elsevier.

7.3.1.3 Rigid TE legs with flexible headers and interconnects

Another design for a TEG uses bulk TE legs with flexible and stretchable interconnect made from a liquid phase, highly conductive alloy called eutectic gallium–indium (EGaIn). This design provides better flexibility, mechanical properties, and reproducibility compared to the previously discussed flexible TEGs [131, 132].

Figure 7.4a shows the fabrication process of such a device. First, a stencil is used to arrange the *p*- and *n*-type bulk legs in a specific order. An elastomer like PDMS is then poured to fill the space between the legs. Next, the EGaIn is deposited to form the top interconnects, and the wires are soldered to the first and last legs. Another layer of elastomer is then added to encapsulate the liquid phase interconnects and the soldered wire. Afterward, the device is flipped, and the interconnects for the bottom of the legs are deposited. The device is then fully encapsulated using an elastomeric material.

Figure 7.4b–d shows actual picture of the device under normal and bending conditions, highlighting its components. As with all TEGs, the performance of this device depends on ambient temperature and air velocity. Lower ambient temperatures and higher air velocity enhance the output power (Figure 7.4e) [131]. After over 1,000 bending cycles, the device maintained its flexibility and performance with only a minimal change in its overall ohmic resistance.

By addressing the thermal conductivity issues at the top and bottom of the device, this method has the potential to replace rigid TEGs while maintaining high power, flexibility, and stretchability. Further improvements can be achieved by incorporating thin copper heat spreaders into the device (inset of Figure 7.4f). When worn on the wrist, this flexible TEG generates an output power density of 5.2 $\mu W/cm^2$ with natural convection at room temperature. This increases to 35 $\mu W/cm^2$ at an air velocity of 1.2 m/s, which is equivalent to an average walking speed (Figure 7.4f) [133].

7.4 Flexible TEGs beyond bulk materials

While methods for fabricating flexible TEGs from bulk materials are popular due to their simplicity and high performance, there is growing interest in developing TEGs beyond rigid materials. This approach is crucial for several reasons: it helps overcome the mechanical limitations of bulk materials, enables better device integration, allows for the use of new TE materials like polymers and halide perovskites, enhances scalability, reduces costs, and opens up new application areas such as flexible Internet of Things (IoT) and smart textiles. This section will discuss various methods for fabricating flexible TEGs that do not rely on rigid materials.

7.4.1 Electrochemical deposition technique

One of the main techniques for fabricating electrical connections in printed circuit boards (PCBs) is electrochemical deposition (ECD). ECD is a thin film fabrication technique in which a material is deposited onto a conductive substrate using an electrochemical reaction. The advantages of this technique include low cost, simplicity, purity, accurate thickness, room-temperature processing, a high growth rate, and the

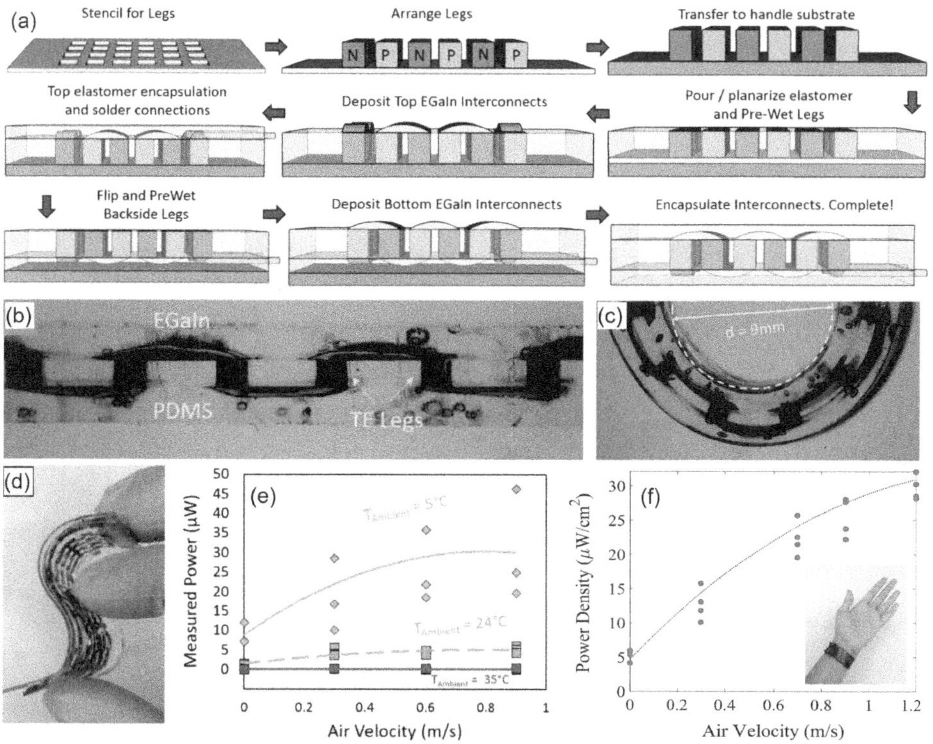

Figure 7.4: (a) Fabrication process for an EGaIn-based flexible TEG using rigid thermoelectric legs. (b) Cross-sectional images showing the TE legs and EGaIn interconnects embedded within the PDMS. (c) Flexible TEG bent to a radius of 9 mm. (d) Photograph of the flexible device. (e) The power density versus air velocities over the device at different ambient temperatures. (f) The power density versus air velocity for an updated EGaIn-based flexible device that incorporates copper headers, longer leg lengths, and a fill fraction of 20% at ambient temperature. The inset shows the device being worn on the wrist. Reproduced from Refs. [129, 131, 133], with permission from Elsevier.

ability to be applied to a wide range of materials while also enabling selective deposition. For fabricating flexible devices, this technique allows for the growth of thin films with a large contact area between the TE legs and the substrate [134]. ECD also enables the fabrication of thin lateral legs on a flexible substrate like PDMS.

The fabrication process for these flexible TEGs involves several steps: initial deposition of TE materials on a silicon substrate, photolithography and deposition of metal contacts, deposition of copper heat spreaders, filling with PDMS, etching and removing of Si and SiO_2 layers, and final packaging into PDMS. Figure 7.5a shows a schematic picture of the device, which contains Bi_2Te_3 and Sb_2Te_3 legs, a copper heat guide, and a PDMS substrate. Figure 7.5b shows an actual image of the flexible device. The inset displays a schematic cross-section of the fabricated device. At an ambient temperature of 15 °C, this device can generate an output power up to 3 μW/cm² [135].

Similar to most thin film TEGs, the main issue limiting the performance of these devices is the contact resistance at the junction between the TE legs and metal interconnects. Additionally, compared to bulk TE materials, Bi_2Te_3 thin films have weaker TE properties and to date they cannot compete with the best bulk materials.

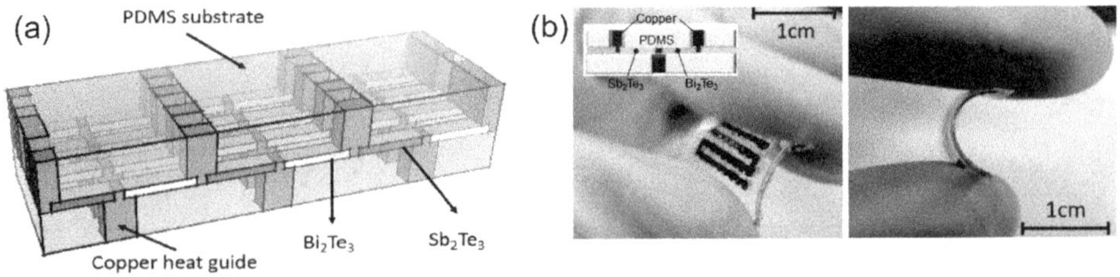

Figure 7.5: (a) Schematic picture of a device fabricated by electrochemical deposition (ECD) technique. (b) Actual picture of the flexible device made from ECD method. The inset shows a schematic picture of the crosses section of the device. Reproduced from Ref. [135], with permission from Elsevier.

7.4.2 Screen printing

Screen printing is a printing or coating technique to create electronic devices. In this method, ink, paste, or slurry of a material is forced through a stencil to make a desired pattern on a substrate. Screen printing is scalable, low-cost, versatile, and reliable for growing thick films on flexible substrates. However, the method has drawbacks, including limited precision in size and thickness, high surface roughness, and reduced uniformity over large surface areas. This technique can be as simple as patterning pastes of *p*- and *n*-type Bi_2Te_3 alloys onto a Kapton tape [136]. Screen-printed flexible TEGs often have high thermal and electrical contact resistance at their junctions and their performance degrades over time. These issues can be mitigated with improved techniques.

Figure 7.6a shows a complete process flow for fabricating a flexible TEG device [137]. This method involves screen printing thick films of *n*-type $Bi_2Se_{0.3}Te_{2.7}$ and *p*-type $Bi_{0.3}Sb_{1.7}Te_3$ along with copper interconnects on the top and bottom of SiO_2/Si/quartz substrates. Then, PDMS is infiltrated to fill the gaps around the TE legs and substrates. The top and bottom quartz substrates are removed by a laser multi-scanning process, and finally the top and bottom SiO_2/Si layers are etched away. Figure 7.6b and c shows a schematic and an actual picture of the flexible TEG fabricated by screen printing. To ensure flexibility and mechanical robustness, the TE legs were grown with a 2-mm diameter and a 650-μm thickness. The flexible TEG fabricated by screen printing can generate 4.78 mW/cm^2 at ΔT of 25 °C. While this power is sufficient to run several sensors at high ΔT values, the output power from wearable devices, which operate at much smaller ΔT values, is still very limited.

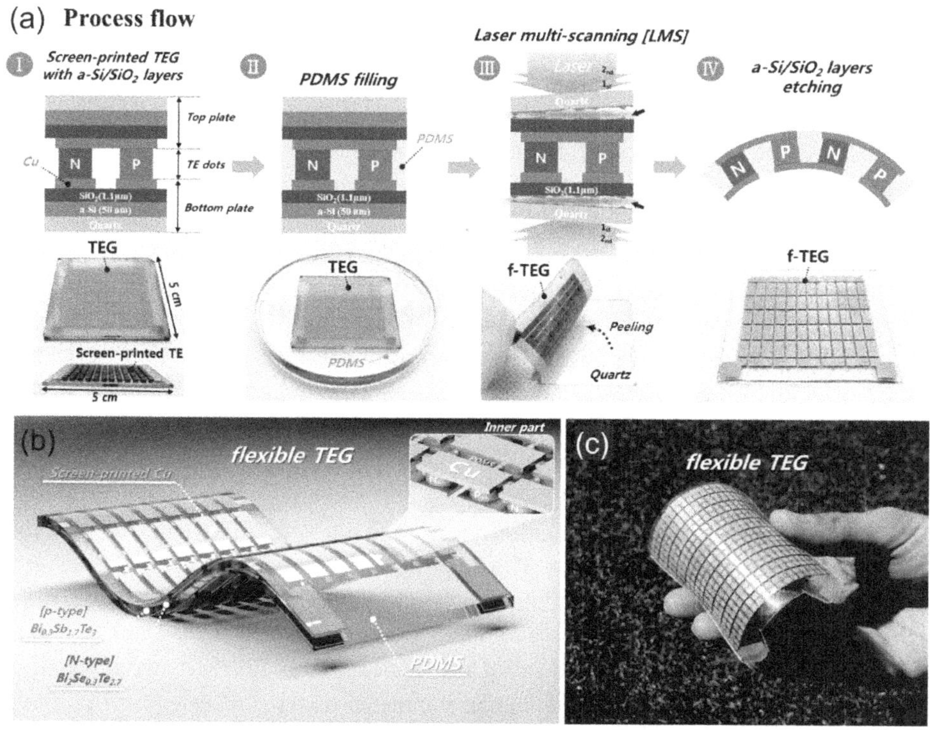

Figure 7.6: (a) Process flow of the screen-printed flexible TEG. (b) Schematic and (c) actual picture of the flexible device. Reproduced with permission from Ref. [137]. Copyright © [2016] American Chemical Society.

7.4.3 Organic materials/solution-based processing

Another method of fabricating flexible TEGs uses organic TE materials and solution-based processing. Traditionally, BiTe-based alloys remain the leading TE materials for applications near room temperature. However, replacing these alloys with a flexible TE material is an attractive and emerging research area.

Poly(3,4-ethylenedioxythiophene), commonly known as PEDOT, which is often combined with polystyrene sulfonate (PSS), is one of the best organic TE materials. It is inherently flexible and has a zT value of around 0.25 [138, 139]. In solution processing, the TE material is either dispersed in a solvent or prepared as ink or paste, which is then coated or cast onto a substrate. The output power from these devices is currently low, and on the order of $n\text{W/cm}^2$ [140]. Further studies are needed to improve the performance of these polymeric flexible devices.

7.4.4 Thermoelectric fabrics

The surface area of human skin is approximately 1.6–2 m^2. When a person wears a short-sleeved shirt and long pants, about 80% of their body is covered with clothing. This makes clothing, with a surface area of about 1.3–1.6 m^2, an ideal platform for wearable technology to generate power for various sensors. For example, a TEG with a very low power density of just 1 $\mu W/cm^2$ can generate up to 16 mW of power, which is sufficient to power several sensors. Therefore, TEGs can be integrated directly into fabrics to generate significant power from body heat.

TE fabrics are flexible and can be woven into clothing to convert body heat into electricity. Researchers have developed different materials for these fabrics, such as alternating stacks of doped carbon nanotube and polyvinylidene fluoride (PVDF) sheets, which offer a low-cost and scalable solution [141]. While other flexible fabrics (like silk) have been explored, the main challenges are high contact resistance and very low TE performance. Among different methods, electrospun polyacrylonitrile (PAN) nanofiber sheets coated with alternating *n*- and *p*-type Bi_2Te_3 and gold interconnect have shown promising results. This approach, which contains high-performance TE materials, low thermal conductivity fabrics, and a high planar density of TE materials, has led to a significant increase in power output [142].

In another work, researchers fabricated a stretchable fabric using alternate *n*-type doped and *p*-type hybridized carbon nanotube fibers (CNTFs). These TE fibers were bent into an arc and pillar shape (Figure 7.7a) [143]. While these structures are highly flexible (Figure 7.7b) and can generate reasonable power at high ΔTs, their output power at ΔT below 10 °C is negligible compared to bulk materials. Furthermore, while integrating bulk TE legs into a flexible fabric could improve performance [144], washability remains a major concern.

The primary challenges for TE fabrics are effectively integrating high-performance TE materials into fibers and reducing contact resistance. Currently, the power generated from TE fabrics is not comparable to that from bulk or other flexible TEGs. However, this field remains promising due to the large surface area that can be covered by fabrics. Therefore, future research must focus on developing new textiles and engineering techniques to achieve a much higher TE performance [145].

7.5 Limitations of flexible TEGs

In addition to the discussed methods for fabricating flexible TEGs, techniques like physical vapor deposition (PVD) [146, 147] and dispenser printing [148, 149] are also used. However, these techniques do not generate high power due to their high electrical contact resistance. As discussed in Chapter 4, TEGs worn on the human body experience a small temperature difference (ΔT) of around 1 °C, which limits the generated power. While recent advancements in flexible TEGs have led to innovations and new

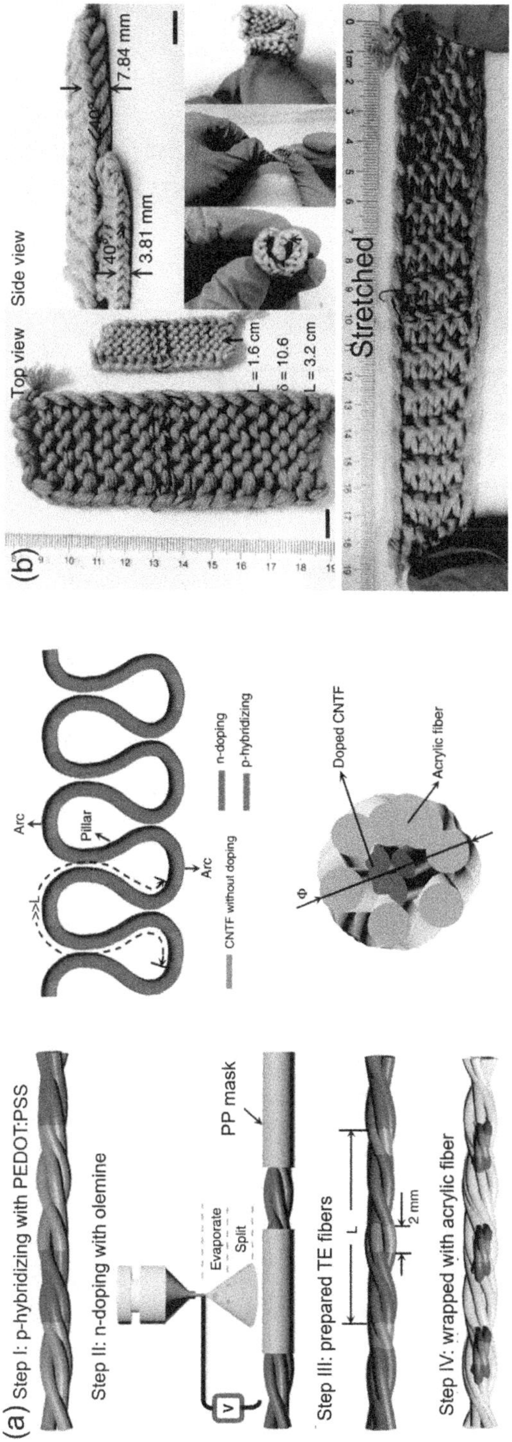

Figure 7.7: (a) Schematic drawing of the fabrication process for a stretchable CNTF woven into a TEG. (b) Actual picture of the woven TEG, demonstrating its different sizes, flexibility, and stretchability. Reproduced from [143] licensed under CC BY 4.0.

fabrication techniques, the output power remains very limited and is not yet comparable to that of bulk TEGs.

Table 7.1 lists the power densities generated by various flexible TEGs. The highest power is, as expected, achieved in devices made from bulk materials. Screen printing also shows promising results. Future research on flexible TEGs, especially those made from thin films, should focus on reducing contact resistances and increasing the mechanical durability of TE legs and metal interconnects.

Table 7.1: Comparison of power density and ΔT of flexible TEGs fabricated using various techniques.

Technique	Year	ΔT (°C)	Power (μW/cm^2)	References
Bulk materials	2017	1	8.75	[131]
Bulk materials	2020	~1	5.2	[133]
Bulk materials	2021	1.7	21.1	[150]
Bulk materials	2023	7.3	7.3	[151]
Screen printing	2016	1	7.3	[137]
Screen printing	2017	10	0.062	[152]
Electrodeposition	2009	2	1	[134]
Fabrics	2012	2	0.001	[141]
Fabrics	2016	10	2	[142]
Fabrics	2016	5	0.004	[153]
Organic solution	2016	3	0.0007	[154]
PVD	2006	2	0.1	[155]
PVD	2011	5	0.0001	[156]

8 Cooling and wearable applications of thermoelectric devices

Beyond body heat harvesting, thermoelectric (TE) devices offer significant potential for localized body cooling and personalized thermal comfort. This capability is ready to substantially expand the daily utility of these devices. While the market for thermoelectric coolers (TECs) is already the largest within the field (valued at over US $700 million), their applications have traditionally been limited to solid-state cooling for electronics, refrigeration, automotive, and telecommunications. However, the use of both body heat harvesters and coolers is rapidly growing in wearable devices and platforms that maintain direct or indirect contact with human skin.

This chapter introduces wearable TECs and their operating mechanisms. Furthermore, the broader applications of TE devices for heating, cooling, and power generation across various platforms will be discussed.

8.1 Wearable thermoelectric coolers

Cooling systems in buildings consume above 20% of the total electricity usage, leading to significant CO_2 emissions from fossil fuels and reliance on unsustainable refrigerants. Traditional cooling methods, which use air convection to cool an entire space, are highly inefficient and dependent on the size and insulation of the building. This high energy consumption and the resulting environmental crisis highlight the need for alternatives. Localized cooling is a practical solution for future smart buildings, and TECs are one of the most promising options. As discussed in Chapter 2, TECs provide solid-state cooling via the Peltier effect. When an electric current is applied, they create a temperature difference, providing a cooling effect. This is the inverse of the Seebeck effect, which is used for power generation and was extensively discussed in previous chapters.

TECs have gained attention in the last few decades because they are compact, silent, lightweight, and have no moving parts or fluids. They are currently used commercially in refrigerators and for precision cooling of microelectronics, such as infrared detectors and laser diodes. These commercial TECs are designed for low thermal resistance environments to achieve a high temperature difference (ΔT) or maximum cooling power. External thermal resistance at the hot and cold junction of the TEC can highly degrade the performance of commercial TECs. Wearable TECs enable the users to feel the body skin temperature locally below ambient, producing a sensation similar to holding an ice cube. However, wearable TECs face different challenges. As mentioned in the discussion of body heat harvesting using TEGs, human skin is a poor conductor of heat. Wearable TECs encounter high thermal contact resistance and low heat loads. They must be specifically designed to work with these conditions, facing

 | https://doi.org/10.1515/9781501514814-008

limitations similar to those of wearable thermal energy harvesting devices. Localized cooling requires significantly less electricity than traditional space cooling, which can lead to substantial energy cost savings. Figure 8.1a and b shows a schematic illustration of a wearable TEC and the expected localized cooling.

For optimal performance, wearable TECs require a specific combination of material properties and device design. In a low thermal resistance environment, like those in electronic cooling, TECs with a high Seebeck coefficient enhance cooling capacity, while high electrical conductivity results in a higher coefficient of performance (COP). In high thermal resistance environments, like those for body cooling, materials with lower thermal conductivity provide a higher COP and better cooling. Commercial TECs, which are used in low thermal resistance environments, have a high fill fraction (FF), typically above 20%. In contrast, devices designed for high thermal resistance environments, such as body cooling, perform better with a low FF, offering up to 1.7 times better cooling performance than commercial devices. For example, experimental results showed that a TEC with a 36% FF and a leg aspect ratio (AR) of 1 provided a $\Delta T = 5.0$ °C on the skin, whereas a TEC with a 5.2% FF and an AR of 1.6 provided a $\Delta T = 8.2$ °C [157]. Figure 8.1c–g shows TECs with different FFs and ARs. These findings confirm the critical importance of material choice and device design for the future of wearable TECs.

This cooling system can be integrated into car heating, ventilation, and air conditioning (HVAC) systems to enhance localized body comfort and reduce energy usage. Future investigations to optimize the design of wearable TECs, specifically through the fabrication of flexible TECs, will broaden their commercial applications. Similar to TEGs, BiTe-based TE materials are currently the best candidates for TECs due to their ease of synthesis, reliable performance, high zT, and stability. They are particularly ideal for applications operating across a temperature range of −50 to 250 °C.

Several commercial devices have recently used TE technology for personalized cooling and heating. For example, Ember Wave wristband provides a cool or warm sensation by either rejecting heat away from or moving it toward the skin (Figure 8.2a) [158]. This localized heating/cooling sensation stimulates skin thermal nociceptors and consequently makes the brain perceive a cooler or warmer environment overall. Another company, Dhama Innovations, produces various wearable garments like Flowtherm heating/cooling vest, wristband and pain relief belt, all of which use TE devices [159]. Additionally, Torras COOLiFY Cyber Neck air conditioner provides rapid cooling/heating via multidirectional airflow around the neck [160].

TECs also have applications in helmets, where they can significantly reduce user discomfort caused by excessive heat and humidity. By actively cooling the helmet's interior and managing moisture levels, TECs dramatically enhance the user's experience, particularly for sport applications [161, 162]. Sony also developed Reon Pocket, a personalized temperature-controlled device designed to locally cool and heat the human body [163]. Furthermore, TE devices are finding utility in gaming controllers

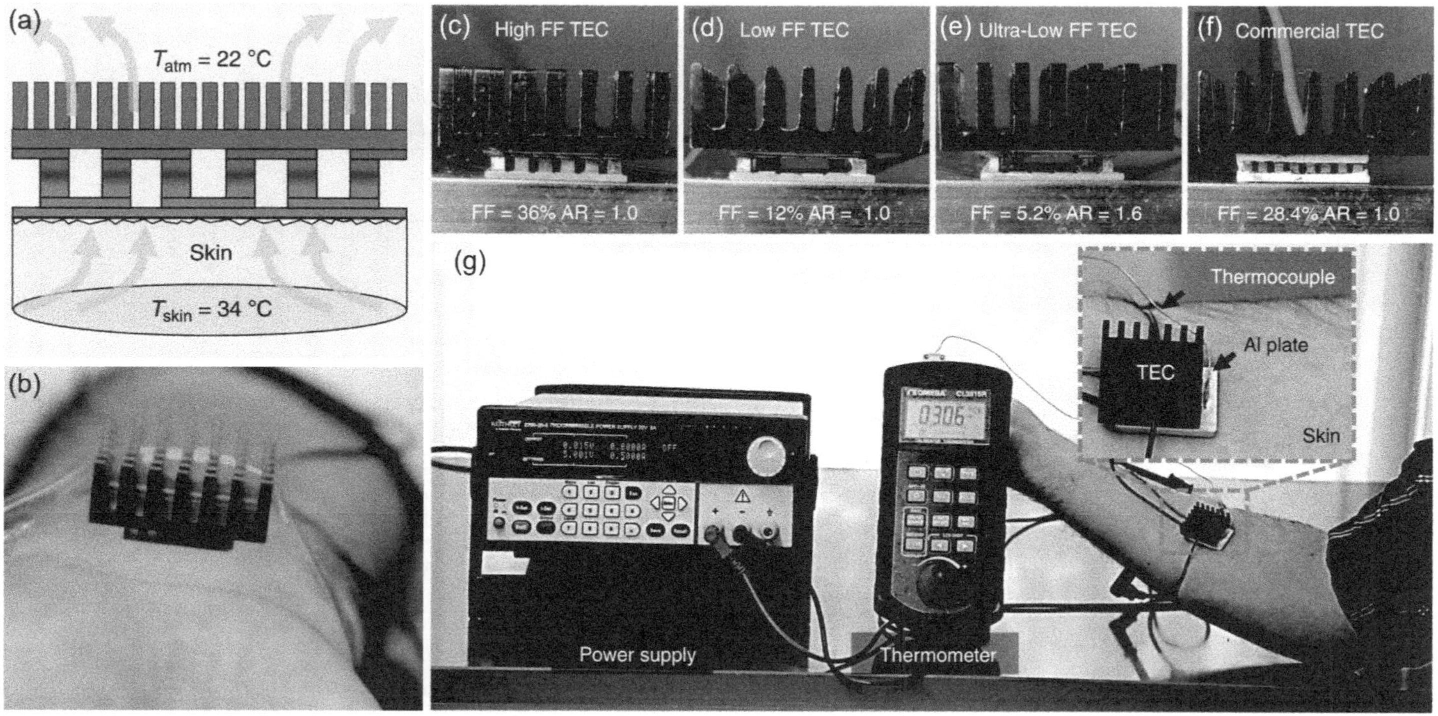

Figure 8.1: (a) Schematic representation of the wearable TEC placed on human skin. (b) A wearable TEC provides localized body cooling. (c–f) TEC modules with different fill fractions (FF) and aspect ratios (AR). (g) Experimental configuration for evaluating the cooling performance of wearable TEC modules. Reproduced from [157] licensed under CC BY 4.0.

and virtual reality (VR) systems. They can enhance immersion by actively mimicking the real environment through heating and cooling the controller based on in-game events.

8.2 Human healthcare and wearables

While the widespread use of TE devices is limited by their low thermal-to-electrical conversion efficiency, their unique properties, such as reliability, compact size, lightweight, and silent, motion-free, and maintenance-free operation, make them suitable for a wide range of applications. These devices also excel at localized heating and cooling, which has opened up extensive uses in human health, sports, military, personal thermal comfort, and entertainment.

The full potential of TE devices is vast, but their performance depends heavily on the properties of the materials used and the specific application. Future innovations in both materials and device designs will undoubtedly broaden their use across various industries. Although several companies have recently focused on innovative TE products, some have been successful while others have been discontinued. It is clear, however, that the field of TEs is experiencing significant growth, particularly in biomedical, wearable, and body-interface applications.

Biomedical application of TEs began in the 1970s, when Medtronic invented the first cardiac pacemaker powered by a radioisotope thermoelectric generator (RTG) [164]. These devices are designed to regulate arrhythmias (irregular heart rhythms) or bradycardia (slow heart rate) by delivering controlled electrical impulses.

The body heat-powered watch stands as one of the pioneered commercial devices to utilize TEGs in wearable technology. In 1998, Seiko® commercialized the first body heat-powered wristwatch. This worn watch was able to generate an open-circuit voltage of 300 mV and 25 μW of power over a 1.5 °C temperature difference (Figure 8.2b) [165]. Following this, Citizen® developed a similar watch in 1999 with an output power of 13.8 μW over a 1 °C temperature difference (Figure 8.2c) [166]. More recently, the Matrix PowerWatch was introduced as the first body heat-powered smartwatch. In addition to displaying time, this smartwatch could wirelessly connect to a smartphone, transmitting data on heart rate, calorie expenditure, step count, and location, all while being entirely powered by its thermoelectric generator (Figure 8.2d) [129].

GreenTEG has commercialized the gSkin, a thermoelectric-powered heat flux sensor designed for core body temperature monitoring (Figure 8.2e) [167]. These sensors monitor the amplitude and phase variations of body temperature to detect physiological abnormalities and assist in health diagnosis. For instance, certain disorders, such as sleep quality issues, burnout, Parkinson's disease, and ovulation, can alter the amplitude of body temperature. In contrast, other conditions, including insomnia, narcolepsy, jet lag, and Alzheimer's disease, result in a phase shift in the body temperature cycle. TE devices can sense temperature variations and/or generate (cold/hot) temperature stimuli to aid in the diagnosis of such diseases.

To enhance athlete performance, Nike, in collaboration with Hyperice, developed the Hyperboot, a TE-embedded recovery boot. This wearable gear utilizes TE devices with accurate sensors to precisely regulate skin temperature, offering instant, optimized heating and cooling during warm-ups and recovery to maintain the range of motion and accelerate muscle preparation [168]. The two companies also developed a complementary vest based on the same technology. Similarly, Dhama Innovations has developed temperature-controlled wearables that provide instant thermal regulation for a wide range of applications. These devices are designed to enhance body comfort, reduce muscle fatigue, and accelerate recovery. They achieve this by assisting the body's natural thermoregulation process, efficiently removing excess heat to cause an immediate, localized drop in temperature [159].

The ChiliPad thermoelectric blanket and mattress are offered for use in body comfort, chemotherapy, or surgical procedures to safely control local or whole-body temperature [169]. Traditional blankets which rely on water circulation or resistive heating elements, pose safety risks, such as the potential for water leakage and/or electrical shock. Similar TE products can also be utilized to thermoregulate patients or individuals experiencing thermal discomfort [170].

Although several commercial TE devices have been introduced for wearable and medical applications, the number of successful applications remains limited to a few companies. However, with continued materials and device innovations, alongside novel design concepts, many new commercial products are expected to be launched in the near future.

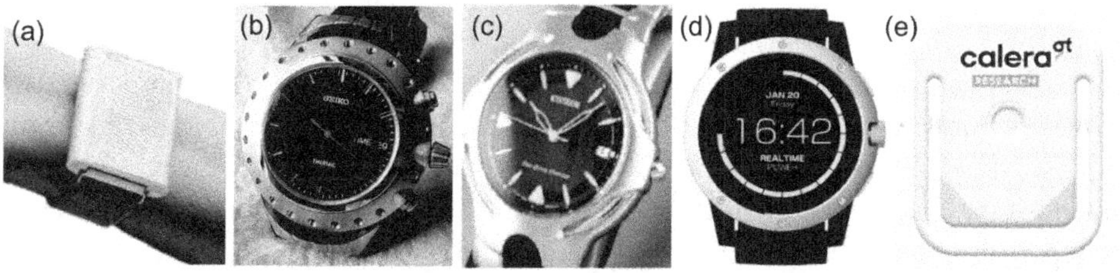

Figure 8.2: Examples of commercial wearable devices: (a) Embr wave, a wearable wristband, uses a TE device to provide a cool or warm sensation. Reproduced from [171] under the Creative Commons CC-BY license. (b) Seiko Thermic watch powered by thermoelectric generator (TEG) can generate ~25 μW from body heat. (c) Citizen Eco-Drive Thermo model watch uses body heat to generate ~13.8 μW power (d) Matrix PowerWatch is the first smartwatch powered by body heat using TEGs ((b), (c), and (d) are adapted from Ref. [129], with permission from Elsevier). (e) gSkin, a non-invasive, thermoelectric-powered body heat flux sensor. Adapted from [172], with permission.

9 Future outlooks and opportunities

The global market for wearable electronics is projected to be a significant and rapidly growing industry. It is anticipated to grow to over USD 500 billion by 2030. This corresponds to a compound annual growth rate of nearly 14–18%, highlighting a massive market that requires extensive research and development from both academia and industry.

Wearable electronics come in many forms, including wristbands, armbands, headbands, hats, helmets, belts, straps, patches, eyeglasses, watches, jewelry, rings, and textiles. These devices are used for wellness, health monitoring, and user comfort. While already used in devices like fitness trackers and smartwatches, self-powered wearables have the potential to significantly advance medical care, patient support, and health monitoring.

Devices that constantly need charging or must remain plugged in face significant functional limitations, hindering their widespread deployment. To overcome the challenge of long-lasting power for electronics, particularly wearable devices, active research worldwide is focusing on three primary solutions: fabricating low-power sensors that require less energy and extend battery life; creating new solid-state batteries that offer extended charges, reducing the need for frequent charging and preventing interruptions in data collection; and deploying standalone devices powered by body energy, and approach that eliminates the reliance on traditional batteries.

While various industries are actively pursuing the first two solutions, the third presents a more practical path forward. The development of battery-free or less battery-dependent wearable devices has become a major interdisciplinary research focus since the early 2010s. Future health monitoring devices will necessitate continuous energy harvesting techniques that can either completely replace current rechargeable batteries or consistently use body energy to charge them over time.

Body energy harvesting for wearable applications is a growing field that requires future research and development in materials, devices, and system design to become a feasible technology. The main body energy harvesting candidates include active methods like piezoelectric, electromagnetic, electrocaloric, and triboelectric generators, which use various concepts to transform kinetic energy into electrical energy, and passive methods such as pyroelectric, photovoltaic, and thermoelectric generators (TEGs), which utilize heat or ambient light to generate electricity. Recently, researchers have also explored biofuel cells, which generate energy by utilizing sweat, glucose, or lactate.

Among the various methods discussed in Chapter 1, thermoelectric (TE) devices are ideal for thermal energy harvesting and thermal management in any application that involves contact with the human body or similar situations with high parasitic resistances. Looking ahead, these technologies could also be applied to localized HVAC systems and smart buildings. The future advancement of wearable TE devices

 | https://doi.org/10.1515/9781501514814-009

is categorized into eight main areas, including improvements in materials, devices, and system design, as well as applications.

(1) Exploration of next-generation thermoelectric materials: The future of TE materials research for wearable applications demands the discovery of a new class of materials. These materials should possess a zT greater than 1.3 and a power factor (PF) exceeding 60 $\mu W/cm \cdot K^2$ at room temperature, coupled with a high Seebeck coefficient and low thermal conductivity. This is crucial for replacing state-of-the-art BiTe-based materials, which appear to have reached their maximum room temperature zT of 1.3. Generally, a high zT is necessary to enhance overall device efficiency and performance, while a high PF is vital for greater output power. Low thermal conductivity is required to match the parasitic thermal resistances in wearable platforms, as discussed in Chapter 4, and to maintain a higher ΔT across the TEG. Finally, a high Seebeck coefficient is essential for generating sufficient output voltage to power an electronic device.

Furthermore, despite recent advancements in TE materials discovery, processing, and device fabrication, most commercial TE devices still rely on the same materials and techniques established decades ago. Therefore, there is a high demand for integrating new materials alongside novel device architectures that can effectively maintain the temperature difference across wearable TEGs. For example, 3D printing is an emerging technique that can eliminate parasitic electrical contact resistance and fabricate devices with unique architectures, including unconventional TE leg sizes and geometries.

(2) Designing high-performance flexible or miniaturized rigid energy harvesters: When it comes to energy harvesting, rigid TEGs currently offer the highest power density among rigid, flexible, and textile options. However, flexibility is a crucial factor for the future of TEGs and wearable devices in general, especially since these devices directly contact the body surface which is flexible. Consider smart buildings, for instance, where TEs could provide localized body heating/cooling or generate power from the human body. For such applications, flexibility is essential for seamless device compatibility with the body. Similarly, smart mattresses, couches, pillows, and cushions will require high-performance flexible TE devices. These devices must be able to deform perfectly and fill all gaps between the body and the device to maximize power generation or cooling efficiency. Therefore, enhancing the performance of flexible TEGs or designing miniaturized rigid TEGs will meet the industrial demand for these devices in wearables, smart buildings, and any other application requiring direct contact with body heat.

(3) Integrating wearables into textiles: Human skin has a vast surface area ranging from 15,000 to 20,000 cm^2. A wearable textile that generates 2 $\mu W/cm^2$, at least ten times less than current flexible TEGs and over 20 times less than rigid ones, if covering even one-tenth of the human skin, could produce over 3 mW of power. This output is sufficient to operate several wearable sensors or conserve energy for alternative usage. This highlights the significant potential of integrating energy harvesters directly into clothing for the future of self-powered wearable devices.

The power output of TE textiles is currently too low, and their fabrication costs are too high. These factors limit their practical application. Integrating bulk or flexible TEGs into textiles and establishing a reliable design could increase their use in clothing. Further research is needed to enhance the TE properties of textiles and integrate them into everyday fibers through methods like printing, spinning, or weaving. It is crucial to investigate optimal temperature gradients, texture architectures, and thermal modeling. Additionally, for widespread application, the durability, stretchability, washability, and overall system design of these textile-based energy harvesters must be thoroughly addressed. Further investigations into implantable and/or skin-integrated TEGs are also highly demanded.

(4) Enhancing mechanical properties and stability: While rigid TEGs generally outperform flexible ones in terms of mechanical properties, understanding and improving these characteristics is crucial for creating durable, high-performance devices. For example, Bi_2Te_3, the state-of-the-art material for body heat harvesting, offers reliable mechanical properties and oxidation resistance. A major challenge for new materials with comparable zT values is their instability and unreliable mechanical properties, which limit their widespread use.

Furthermore, as discussed in Chapter 4, devices with a low fill fraction can generate more power, but their potential for mechanical failure can compromise performance. Thinner and longer TE legs are also ideal for wearable TEGs, but fabricating and dicing them requires superior mechanical properties that many semiconductors lack.

While rigid TE materials are brittle, the mechanical load is not typically applied directly to the materials or electronic junctions, which offers some protection. However, flexible devices are more prone to failure because the mechanical load may be applied directly to the TE legs or junctions. Therefore, it is essential to perform various mechanical tests, such as fatigue, twisting, bending, and stretching, on flexible devices to fully characterize their behavior for real-world applications.

(5) Improving the performance and efficiency of DC–DC boost converters: Although the efficiency of DC–DC boost converters has improved tremendously in recent years, their performance on wearable systems still needs to be upgraded. This is especially true for TEGs, which produce voltages in the millivolt range. Furthermore, the size and flexibility of these converters must be adjusted to match the advancements in miniaturized and flexible TEGs. Future research should aim to improve their efficiency, power density, and reliability. Additionally, there is a demand for developing intelligent converters with smart control features that can adapt to dynamic loads and varied conditions.

(6) Wearable coolers: Wearable thermoelectric coolers (TECs) for body comfort and HVAC systems have gained recent attention due to the psychological effect that warm and cold sensations have on a person's perception of temperature. The performance of these wearable coolers, however, is subject to the same limitations as other wearable TEs. In particular, most commercial devices are not optimized for personalized body cooling. Therefore, advancements in TE materials, devices, and systems are needed to enable high-performance, low-power coolers that work efficiently in contact with body.

(7) Body heat harvesting from animals: Self-powered animal wearables represent an emerging technology for smart and precision livestock farming, ensuring the health and welfare of animals. Since daily rechargeable batteries are not a viable option for large populations, these self-powered devices offer a sustainable solution. Harvesting animal body heat using TEGs can provide enough energy to continuously power multiple health monitoring sensors.

These wearables, such as smart collars, enable continuous real-time tracking of physiological parameters including heart rate, body temperature, and activity, facilitating the early detection of illness or stress. The autonomous operation of these systems reduces labor and maintenance requirements, allowing deployment across large populations and consequently lowering operational costs. Continuous health monitoring further improves livestock productivity and overall farm efficiency, while timely intervention can reduce veterinary expenses and mitigate disease transmission. Additionally, integrated tracking capabilities support behavioral analysis, optimizing feeding, and reproduction management. Self-powered wearables thus represent a promising avenue for advancing sustainable and data-driven livestock management.

(8) Novel applications: Prior to 2010, the applications for TE materials and devices were limited, but they have since expanded significantly. While many of these applications feature ideal contact with minimal thermal resistance, others face high thermal resistance at the junctions of the heat source and sink. To overcome this, there is a need to extend the use of TE devices to platforms that are in direct contact with the human body. These applications include electronic handheld devices such as game controllers, tools, medical diagnostic instruments, and barcode readers; seating platforms like office chairs, armchairs, couches, and dental units; lying and resting platforms such as beds, mattresses, therapy tables, and sleeping mats; mobile and supportive platforms including wheelchairs and seats in cars, bikes, airplanes, and trains; and floor-contact platforms like floor cushions and yoga mats.

The power generated by wearable TEGs depends on the skin temperature of the user, which varies due to metabolic differences. Therefore, future research should report the age, gender, and weight of subjects. Additionally, to collect more realistic and comparable data, it is recommended to test TEGs on a diverse range of body types.

Body energy harvesters, while not yet capable of fully replacing batteries, can significantly extend battery life and reduce the need for frequent charging. In addition to the research directions mentioned above, truly self-powered wearable devices might be achieved through the development of ultra-low-power electronic sensors.

Despite the rapid growth of the health monitoring device market, several challenges remain. Discontinuous data collection, inconsistent results between different devices, and a lack of proper user training for data acquisition and interpretation can all impact their accuracy, reliability, and reputation. To address these issues, establishing self-powered and standardized equipment along with proper data handling protocols is crucial.

Abbreviations

3D	Three-dimensional
AFE	Analog front-end
Al_2O_3	Alumina
AlN	Aluminum nitride
AR	Aspect ratio
$BaTiO_3$	Barium titanate
BiTe	Bismuth telluride
CAGR	Compound annual growth rate
CCD	Charge-coupled device
CNTF	Carbon nanotube fibers
COP	Coefficient of performance
COTS	Commercial off-the-shelf
CPU	Central processing unit
d_{31}	Transverse mode
d_{33}	Longitudinal mode
DBC	Direct-bonded copper
DC	Direct current
ECD	Electrochemical deposition
ECG	Electrocardiogram
EEG	Electroencephalography
E_F	Fermi energy
EGaIn	Eutectic gallium-indium
EMG	Electromyogram
ESG	Electrostatic generator
FAST	Field-assisted sintering technique
FF	Fill fraction
hH	Half-Heusler
HVAC	Heating, ventilation, and air conditioning
IoT	Internet of Things
KNN	Potassium sodium niobate
LED	Light-emitting diodes
$LiTaO_3$	Lithium tantalate
Li-ion	Lithium ion
Li-Po	Lithium polymer
MEMS	Microelectromechanical systems
MIG	Magnetic induction generator
NASA	National Aeronautics and Space Administration
PAN	Polyacrylonitrile
PANI	Polyaniline
PCB	Printed circuit board
PDMS	Polydimethylsiloxane
PEDOT	Poly (3,4-ethylenedioxythiophene)
PEG	Pyroelectric generator
PF	Power factor
PMN-PT	Lead magnesium niobate–lead titanate
PMU	Power management unit
PP	Polypropylene

 | https://doi.org/10.1515/9781501514814-010

PPG	Photoplethysmography
PPy	Polypyrrole
PSS	Polystyrene sulfonate
PVC	Polyvinyl chloride
PVDF	Polyvinylidene fluoride
PVG	Photovoltaic generator
PZG	Piezoelectric generator
PZT	Lead zirconate titanate
RTG	Radioisotope thermoelectric generator
SC	Solar cell
SHS	Self-propagating high-temperature synthesis
SoC	System-on-chip
SPS	Spark plasma sintering
TE	Thermoelectric
TEC	Thermoelectric cooler
TEG	Thermoelectric generator
TENG	Triboelectric nanogenerator
T_C	Cold-side temperature
TG	Triboelectric generator
T_H	Hot-side temperature
VR	Virtual reality
ZnO	Zinc oxide
zT	Thermoelectric figure of merit

References

[1] A. Nozariasbmarz, J. H. Dycus, M. J. Cabral, C. M. Flack, J. S. Krasinski, J. M. LeBeau, D. Vashaee, Efficient self-powered wearable electronic systems enabled by microwave processed thermoelectric materials, Appl. Energy, 283, 116211, (2021).

[2] G. D. Szarka, B. H. Stark, S. G. Burrow, Review of power conditioning for kinetic energy harvesting systems, IEEE Trans. Power Electron., 27(2), 803–815, (2012).

[3] S. Priya, Advances in energy harvesting using low profile piezoelectric transducers, J. Electroceram., 19, 165–182, (2007).

[4] A. Toprak, O. Tigli, Piezoelectric energy harvesting: State-of-the-art and challenges, Appl. Phys. Rev., 1(3), 031104, (2014).

[5] S. Priya, D. J. Inman, Energy harvesting technologies, Springer, Boston, MA, (2009).

[6] M. Zhou, M. S. H. Al-Furjan, J. Zou, W. Liu, A review on heat and mechanical energy harvesting from human – Principles, prototypes and perspective, Renew. Sustain. Energy Rev., 82(3), 3582–3609, (2017).

[7] C. Dagdeviren, Z. Li, Z. L. Wang, Energy harvesting from the animal human body for self-powered electronics, Annu. Rev. Biomed. Eng., 19, 85–108, (2017).

[8] Y. Liu, H. Khanbareh, M. A. Halim, A. Feeney, X. Zhang, H. Heidari, R. Ghannam, Piezoelectric energy harvesting for self-powered wearable upper limb applications, Nano Sel., 2, 1459–1479, (2021).

[9] E. Bouendeu, A. Greiner, P. J. Smith, J. G. Korvink, A low-cost electromagnetic generator for vibration energy harvesting, IEEE Sens. J., 11(1), 107–113, (2011).

[10] D. F. Berdy, D. J. Valentino, D. Peroulis, Kinetic energy harvesting from human walking and running using a magnetic levitation energy harvester, Sens. Actuators A Phys., 222, 262–271, (2015).

[11] W. Lin, Y. Wei, X. Wang, K. Zhai, X. Ji, Study on human motion energy harvesting devices: A review, Machines, 11, 977, (2023).

[12] M. Hayakawa, A study of the new energy system for quartz watches (II). The effective circuit for the system, Proc. Congres Europeen de Chronométrie (C.E.C), 61–85, (1988).

[13] R. Amirtharajah, A. Chandrakasan, Self-powered signal processing using vibration-based power generation, IEEE J. Solid-State Circuits, 33(5), 687–695, (1998).

[14] J. Y. Hayashida, Unobtrusive integration of magnetic generator systems into common footwear, Master's thesis, Massachusetts Institute of Technology, (2000).

[15] H. Goto, T. Sugiura, Y. Harada, T. Kazui, Feasibility of using the automatic generating system for quartz watches as a leadless pacemaker power source, Med. Biol. Eng. Comput., 37(1), 377–380, (1999).

[16] T. von Büren, P. Mitcheson, T. Green, E. Yeatman, A. Holmes, G. Troster, Optimization of inertial micropower generators for human walking motion, IEEE Sens. J., 6(1), 28–38, (2006).

[17] C. R. Saha, T. O'Donnell, N. Wang, P. McCloskey, Electromagnetic generator for harvesting energy from human motion, Sens. Actuators A Phys., 147(1), 248–253, (2008).

[18] F. U. Khan, M. U. Qadir, State-of-the-art in vibration-based electrostatic energy harvesting, J. Micromech. Microeng., 26, 103001, (2016).

[19] R. T. Aljadiri, L. Y. Taha, P. Ivey, Electrostatic energy harvesting systems: A better understanding of their sustainability, J. Clean Energy Technol., 5(5), 409–416, (2017).

[20] S. Boisseau, G. Despesse, Seddik, Electrostatic conversion for vibration energy harvesting, Small-Scale Energy Harvest., (2012).

[21] T. Ghomian, S. Mehraeen, Survey of energy scavenging for wearable and implantable devices, Energy, 178, 33–49, (2019).

[22] F. Wang, O. Hansen, Electrostatic energy harvesting device with out-of-the-plane gap closing scheme, Sens. Actuators A Phys., 211, 131–137, (2014).

[23] L. Bu, H. Xu, B. Xu, L. Song, Micro-fabricated liquid encapsulated energy harvester with polymer barrier layer as liquid electret interface, J. Phys. Conf. Ser., 557, 0120361, (2014).

 | https://doi.org/10.1515/9781501514814-011

[24] C. Lagomarsini, A. Kachroudi, S. Basrour, C. J. Mistral, A. Sylvestre, Autonomous electrostatic generator for energy harvesting applications under inertial load, Proc. SPIE, 10594(105941Q), (2018).
[25] M. Lallart, S. Pruvost, D. Guyomar, Electrostatic energy harvesting enhancement using variable equivalent permittivity, Phys. Lett. A, 375(45), 3921–3924, (2011).
[26] M. Ha, J. Park, Y. Lee, H. Ko, Triboelectric generators and sensors for self-powered wearable electronics, ACS Nano, 9, 3421–3427, (2015).
[27] Z. L. Wang, Triboelectric nanogenerators as new energy technology for self-powered systems and as active mechanical and chemical sensors, ACS Nano, 7(11), 9533–9557, (2013).
[28] F.-R. Fan, Z.-Q. Tian, Z. L. Wang, Flexible triboelectric generator, Nano Energy, 1, 328–334, (2012).
[29] X. Pu, M. Liu, X. Chen, J. Sun, C. Du, Y. Zhang, J. Zhai, W. Hu, Z. L. Wang, Ultrastretchable, transparent triboelectric nanogenerator as electronic skin for biomechanical energy harvesting and tactile sensing, Sci. Adv., 3(5), e1700015, (2017).
[30] K. Dong, J. Deng, Y. Zi, Y.-C. Wang, C. Xu, H. Zou, W. Ding, Y. Dai, B. Gu, B. Sun, Z. L. Wang, 3D orthogonal woven triboelectric nanogenerator for effective biomechanical energy harvesting and as self-powered active motion sensors, Adv. Mater., 29(38), 1702648, (2017).
[31] R. Zhang, J. Örtegren, M. Hummelgård, M. Olsen, H. Andersson, H. Olin, Harvesting triboelectricity from the human body using non-electrode triboelectric nanogenerators, Nano Energy, 45, 298–303, (2018).
[32] G. Khandelwal, R. Dahiya, Self-powered active sensing based on triboelectric generators, Adv. Mater., 34, 2200724, (2022).
[33] C. R. Bowen, J. Taylor, E. LeBoulbar, D. Zabek, A. Chauhan, R. Vaish, Pyroelectric materials and devices for energy harvesting applications, Energy Environ. Sci., 7(12), 3817–4148, (2014).
[34] A. Navid, L. Pilon, Pyroelectric energy harvesting using Olsen cycles in purified and porous poly (vinylidene fluoride-trifluoroethylene) [p(VDF-TrFE)] thin films, Smart Mater. Struct., 20(2), 025012, (2011).
[35] S. K. T. Ravindran, T. Huesgen, M. Kroener, P. Woias, A self-sustaining micro thermomechanic-pyroelectric generator, Appl. Phys. Lett., 99(10), 104102, (2011).
[36] A. Potnuru, Y. Tadesse, Characterization of pyroelectric materials for energy harvesting from human body, Integr. Ferroelectr, 150(1), 23–50, (2014).
[37] A. Navid, L. Pilon, Pyroelectric energy harvesting using Olsen cycles in purified and porous poly (vinylidene fluoride-trifluoroethylene) [p(VDF-TrFE)] thin films, Smart Mater. Struct., 20(2), 025012, (2011).
[38] A. Sultana, M. M. Alam, T. R. Middya, D. Mandal, A pyroelectric generator as a self-powered temperature sensor for sustainable thermal energy harvesting from waste heat and human body heat, Appl. Energy, 221, 299–307, (2018).
[39] T. Soga, Nanostructured materials for solar energy conversion, 1st Edition, Netherlands, (2006).
[40] G. K. Grandhi, G. Koutsourakis, J. C. Blakesley, F. De Rossi, F. Brunetti, S. Öz, A. Sinicropi, M. L. Parisi, T. M. Brown, M. J. Carnie, R. L. Z. Hoye, P. Vivo, Promises and challenges of indoor photovoltaics, Nat. Rev. Clean Technol., 1, 132–147, (2025).
[41] X. Hu, Z. Huang, X. Zhou, P. Li, Y. Wang, Z. Huang, M. Su, W. Ren, F. Li, M. Li, Y. Chen, Y. Song, Wearable large-scale perovskite solar-power source via nanocellular scaffold, Adv. Mater., 29, 1703236, (2017).
[42] J. F. Geisz, R. M. France, K. L. Schulte, M. A. Steiner, A. G. Norman, H. L. Guthrey, M. R. Young, T. Song, T. Moriarty, Six-junction III–V solar cells with 47.1% conversion efficiency under 143 Suns concentration, Nat. Energy, 5(4), 326–335, (2020).
[43] J. Fu, P. W. K. Fong, H. Liu, C.-S. Huang, X. Lu, S. Lu, M. Abdelsamie, T. Kodalle, C. M. Sutter-Fella, Y. Yang, G. Li, 19.31% binary organic solar cell and low non-radiative recombination enabled by non-monotonic intermediate state transition, Nat. Commun., 14, 1760, (2023).
[44] A. Chakraborty, G. Lucarelli, J. Xu, Z. Skafi, S. Castro-Hermosa, A. B. Kaveramma, R. G. Balakrishna, T. M. Brown, Photovoltaics for indoor energy harvesting, Nano Energy, 128, 109932, (2024).

[45] J. P. Dieffenderfer, E. Beppler, T. Novak, E. Whitmire, R. Jayakumar, C. Randall, W. Qu, R. Rajagopalan, A. Bozkurt, "Solar powered wrist worn acquisition system for continuous photoplethysmogram monitoring", in: Proceedings of the 36th Annual International Conference of the IEEE Engineering in Medicine and Biology Society (EMBC 2014), Chicago, IL, (2014).
[46] S. Pan, Z. Yang, P. Chen, J. Deng, H. Li, H. Peng, Wearable solar cells by stacking textile electrodes, Angew. Chem. Int. Ed., 53, 6110–6114, (2014).
[47] S. A. Hashemi, S. Ramakrishna, A. G. Aberle, Recent progress in flexible–wearable solar cells for self-powered electronic devices, Energy Environ. Sci., 13, 685–743, (2020).
[48] B. J. Kim, D. H. Kim, Y.-Y. Lee, H.-W. Shin, G. S. Han, J. S. Hong, K. Mahmood, T. K. Ahn, Y.-C. Joo, K. S. Hong, N.-G. Park, S. Lee, H. S. Jung, Highly efficient and bending durable perovskite solar cells: Toward a wearable power source, Energy Environ. Sci., 8, 916–921, (2015).
[49] J. Wan, R. Zhang, Y. Li, Y. Li, Applications of organic solar cells in wearable electronics, Wearable Electron., 1, 26–40, (2024).
[50] G. S. Nolas, J. Sharp, H. J. Goldsmid, Thermoelectrics: Basic principles and new materials developments, Springer, Berlin, (2001).
[51] A. Nozariasbmarz, F. Suarez, J. Dycus, M. Cabral, J. Lebeau, M. C. Öztürk, D. Vashaee, Thermoelectric generators for wearable body heat harvesting: material and device concurrent optimization, Nano Energy, 67, 104265, (2020).
[52] H. G. Yeo, T. Xue, S. Roundy, X. Ma, C. Rahn, S. Trolier-McKinstry, Strongly (001) oriented bimorph PZT film on metal foils grown by RF-sputtering for wrist-worn piezoelectric energy harvesters, Adv. Funct. Mater, 28, 1801327, (2018).
[53] R. Morais, N. Silva, P. Santos, C. Frias, J. Ferreira, A. Ramos, J. Simõesd, J. Baptista, M. Reis, Permanent magnet vibration power generator as an embedded mechanism for smart hip prosthesis, Procedia Eng., 5, 766–769, (2010).
[54] Y. Chiu, V. F. Tseng, A capacitive vibration-to-electricity energy converter with integrated mechanical switches, J. Micromech. Microeng., 18(10), 104004, (2008).
[55] G. Zhu, Y. S. Zhou, P. Bai, X. S. Meng, Q. Jing, J. Chen, Z. L. Wang, A shape-adaptive thin-film-based approach for 50% high-efficiency energy generation through micro-grating sliding electrification, Adv. Mater., 26(23), 3788–3796, (2014).
[56] A. Potnuru, Y. Tadesse, Characterization of pyroelectric materials for energy harvesting from human body, Integr. Ferroelectr, 150(1), 23–50, (2014).
[57] T. Ghomian, O. Kizilkaya, J.-W. Choi, Lead sulfide colloidal quantum dot photovoltaic cell for energy harvesting from human body thermal radiation, Appl. Energy, 230, 761–768, (2018).
[58] G. J. Snyder, E. S. Toberer, Complex thermoelectric materials, Nat. Mater., 7, 105–114, (2008).
[59] S. M. Pourkiaei et al. Thermoelectric cooler and thermoelectric generator devices: A review of present and potential applications, modeling and materials, Energy, 186, 115849, (2019).
[60] T. M. Tritt, Thermoelectric phenomena, materials, and applications, Annu. Rev. Mater. Res., 41, 433–448, (2011).
[61] T. J. Seebeck, A. K. Akad, Wiss, Berlin, 265, (1823).
[62] G. A. Slack, D. M. Rowe, Thermoelectric handbook, Edition. Rowe DM. Chemical Rubber Company, Boca Raton, FL, 407, (1995).
[63] H. J. Goldsmid, Applications of thermoelectricity, Methuen, London, (1960).
[64] J. Heremans, Nanometer-scale thermoelectric materials, Springer Handbook of Nanotechnology, 345–374, (2007).
[65] Y. Lan et al. Enhancement of thermoelectric figure-of-merit by a bulk nanostructuring approach, Advanced Functional Materials, 20(3), 357–376, (2010).
[66] E. A. Liberman, On the way to a new science, Biosystems, 215–216, 104647, (2022).
[67] https://science.nasa.gov/gallery/images-of-voyager/ (Available October 2025).
[68] C. Vining, An inconvenient truth about thermoelectrics, Nat. Mater., 8, 83–85, (2009).

[69] B. Orr, A. Akbarzadeh, M. Mochizuki, R. Singh, A review of car waste heat recovery systems utilising thermoelectric generators and heat pipes, Appl. Therm. Eng., 101, 490–495, (2016).
[70] A. Nozariasbmarz, R. A. Kishore, B. Poudel, U. Saparamadu, W. Li, R. Cruz, S. Priya, High power density body heat energy harvesting, ACS Appl. Mater. Interfaces, 11(43), 40107–40113, (2019).
[71] L. I. Anatychuk, O. J. Luste, Physical principles of microminiaturization in thermoelectricity, in: 15th international conference on thermoelectrics, IEEE, 279–287, (1996).
[72] M. S. El-Genk, H. H. Saber, CRC handbook of thermoelectrics: Micro to nano, D. M. Rowe (Ed), CRC Press, Boca Raton, FL, USA, Ch. 43, (2006).
[73] A. Nozariasbmarz, B. Poudel, W. Li, H. B. Kang, H. Zhu, S. Priya, Bismuth telluride thermoelectrics with 8% module efficiency for waste heat recovery application, iScience, 23(7), 101340, (2020).
[74] P. Webb, Eur. J. Appl. Phys iol. Occup. Physiol., 64, 471–476, (1992).
[75] F. Suarez, A. Nozariasbmarz, D. Vashaee, M. C. Öztürk, Designing thermoelectric generators for self-powered wearable electronics, Energy Environ. Sci., 9, 2099–2113, (2016).
[76] R. A. Kishore, A. Nozariasbmarz, B. Poudel, M. Sanghadasa, S. Priya, Ultra high-performance thermoelectric generators for field deployments, ACS Appl. Mater. Interfaces, 12(9), 10389–10401, (2020).
[77] M. Lossec, B. Multon, H. Ben Ahmed, C. Goupil,Thermoelectric generator placed on the human body: system modeling and energy conversion improvements, Eur. Phys. J. Appl. Phys., 52, 11103, (2010).
[78] T. L. Bergman, F. P. Incropera, D. P. DeWitt, A. S. Lavine, Fundamentals of heat and mass transfer, John Wiley & Sons, (2011).
[79] P. Teertstra, M. Yovanovich, J. Culham, Analytical forced convection modeling of plate fin heat sinks, J. Electron. Manuf., 10, 253–261, (2000).
[80] A. Montecucco, J. Siviter, A. R. Knox, Constant heat characterisation and geometrical optimisation of thermoelectric generators, Appl. Energy, 149, 248–258, (2015).
[81] H. Zhu, W. Li, A. Nozariasbmarz, N. Liu, Y. Zhang, B. Poudel, S. Priya, Half-Heusler alloys as emerging high power density thermoelectric cooling materials, Nat. Commun., 14(1), 3300, (2023).
[82] P. Mayer, R. Ram, Optimization of heat sink limited thermoelectric generators, Nanoscale Microscale Thermophys. Eng., 10, 143–155, (2006).
[83] A. Nozariasbmarz, R. A. Kishore, W. Li, Y. Zhang, L. Zheng, M. Sanghadasa, B. Poudel, S. Priya, Thermoelectric coolers for high-power-density 3D electronics heat management, Appl. Phys. Lett., 120, 164101, (2022).
[84] H. S. Lee, Optimal design of thermoelectric devices with dimensional analysis, Appl. Energy, 106, 79–88, (2013).
[85] W. Liu, Q. Jie, H. S. Kim, Z. Ren, Current progress and future challenges in thermoelectric power generation: From materials to devices, Acta Mater., 87, 357–376, (2015).
[86] M. Gao, Thermoelectric module design theories, in: D. M. Rowe (Ed.), CRC handbook of thermoelectrics: Micro to nano, CRC Press, Boca Raton, FL, (2006).
[87] D. Ebling, K. Bartholome, M. Bartel, M. Jagle, Module geometry and contact resistance of thermoelectric generators analyzed by multiphysics simulation, J. Electron. Mater., 39, 1376, (2010).
[88] T. Nemoto, T. Iida, J. Sato, T. Sakamoto, T. Nakajima, Y. Takanashi, Power generation characteristics of Mg2Si uni-leg thermoelectric generator, J. Electron. Mater., 41, 1312–1316, (2012).
[89] H. S. Kim, T. Itoh, T. Iida, M. Taya, K. Kikuchi, Design of linear shaped thermoelectric generator and self-integration using shape memory alloy, Mater. Sci. Eng. B, 183, 61–68, (2014).
[90] B. Poudel, Q. Hao, Y. Ma, Y. Lan, A. Minnich, B. Yu, X. Yan, D. Wang, A. Muto, D. Vashaee, X. Chen, J. Liu, M. S. Dresselhaus, G. Chen, Z. Ren, High-thermoelectric performance of nanostructured bismuth antimony telluride bulk alloys, Science, 320, 634–638, (2008).
[91] Z. Xu, H. Wu, T. Zhu, C. Fu, X. Liu, L. Hu, J. He, J. He, X. Zhao, Attaining high mid-temperature performance in (Bi,Sb)2Te3 thermoelectric materials via synergistic optimization, NPG Asia Mater., 8, e302, (2016).

[92] S. I. Kim, K. H. Lee, H. A. Mun, H. S. Kim, S. W. Hwang, J. W. Roh, D. J. Yang, W. H. Shin, X. S. Li, Y. H. Lee, G. J. Snyder, S. W. Kim, Dense dislocation arrays embedded in grain boundaries for high-performance bulk thermoelectrics, Science, 348, 109–114, (2015).

[93] Y. Zheng, Q. Zhang, X. Su, H. Xie, S. Shu, T. Chen, G. Tan, Y. Yan, X. Tang, C. Uher, G. J. Snyder, Mechanically robust BiSbTe alloys with superior thermoelectric performance: A case study of stable hierarchical nanostructured thermoelectric materials, Adv. Energy Mater., 5, 1401391, (2015).

[94] W. Xie, X. Tang, Y. Yan, Q. Zhang, T. M. Tritt, Unique nanostructures and enhanced thermoelectric performance of melt-spun BiSbTe alloys, Appl. Phys. Lett., 94, 102111, (2009).

[95] G. Zheng, X. Su, H. Xie, Y. Shu, T. Liang, X. She, W. Liu, Y. Yan, Q. Zhang, C. Uher, M. G. Kanatzidis, X. Tang, High thermoelectric performance of p-BiSbTe compounds prepared by ultra-fast thermally induced reaction, Energy Environ. Sci., 10, 2638, (2017).

[96] Y. Pan, T. Wei, Q. Cao, J. Li, Mechanically enhanced p- and n-type Bi2Te3-based thermoelectric materials reprocessed from commercial ingots by ball milling and spark plasma sintering, Mater. Sci. Eng. B, 197, 75–81, (2015).

[97] S. Wang, G. Tan, W. Xie, G. Zheng, H. Li, J. Yang, X. Tang, Enhanced thermoelectric properties of Bi2 (Te1_xSex)3-based compounds as n-type legs for low-temperature power generation, J. Mater. Chem., 22, 20943–20951, (2012).

[98] X. Yan, B. Poudel, Y. Ma, W. S. Liu, G. Joshi, H. Wang, Y. C. Lan, D. Z. Wang, G. Chen, Z. F. Ren, Experimental studies on anisotropic thermoelectric properties and structures of n-type Bi2Te2.7Se0.3, Nano Lett., 10, 3373, (2010).

[99] W.-S. Liu, Q. Zhang, Y. Lan, S. Chen, X. Yan, Q. Zhang, H. Wang, D. Wang, G. Chen, Z. Ren, Thermoelectric property studies on Cu-doped n-type CuxBi2Te2.7Se0.3 nanocomposites, Adv. Energy Mater., XX, 1–11, (2011).

[100] S. Song, J. Wang, B. Xu, X. Lei, H. Jiang, Y. Jin, Q. Zhang, Z. Ren, Thermoelectric properties of n-type Bi2Te2.7Se0.3 with addition of nano-ZnO:Al particles, Mater. Res. Express, 1, 035901, (2014).

[101] M. Hong, T. C. Chasapis, Z. Chen, L. Yang, M. G. Kanatzidis, G. J. Snyder, J. Zou, n-Type Bi2Te3-xSex nanoplates with enhanced thermoelectric efficiency driven by wide frequency phonon scatterings and synergistic carrier scatterings, ACS Nano, 10, 4719–4727, (2016).

[102] L. Hu, T. Zhu, X. Liu, X. Zhao, Point defect engineering of high-performance bismuth-telluride-based thermoelectric materials, Adv. Funct. Mater, 24, 5211–5218, (2014).

[103] W. Liu, K. C. Lukas, K. McEnaney, S. Lee, Q. Zhang, C. P. Opeil, G. Chen, Z. Ren, Studies on the Bi2Te3–Bi2Se3–Bi2S3 system for mid-temperature thermoelectric energy conversion, Energy Environ. Sci., 6, 552–560, (2013).

[104] A. Nozariasbmarz, M. Hosseini, D. Vashaee, Interfacial ponderomotive force in solids leads to field induced dissolution of materials and formation of non-equilibrium nanocomposites, Acta Mater., 179, 85–92, (2019).

[105] J. Mao, H. Zhu, Z. Ding, Z. Liu, G. A. Gamage, G. Chen, Z. Ren, High thermoelectric cooling performance of n-type Mg3Bi2-based materials, Science, 365, 495, (2019).

[106] F. Zhang, C. Chen, H. Yao, F. Bai, L. Yin, X. Li, S. Li, W. Xue, Y. Wang, F. Cao, X. Liu, J. Sui, Q. Zhang, High-performance N-type Mg3Sb2 towards thermoelectric application near room temperature, Adv. Funct. Mater, 30, 1906143, (2020).

[107] L. Zheng, A. Nozariasbmarz, Y. Hou, J. Yoon, W. Li, Y. Zhang, H. Wu, D. Yang, T. Ye, M. Sanghadasa, B. Poudel, S. Priya, K. Wang, A universal all-solid synthesis for high throughput production of halide perovskite, Nat. Commun., 13, 7399, (2022).

[108] E. J. Carlson, K. Strunz, B. P. Otis, A 20 mV input boost converter with efficient digital control for thermoelectric energy harvesting, IEEE J. Solid-State Circuits, 45(4), 741–750, (2010).

[109] J.-P. Im, S.-W. Wang, K.-H. Lee, Y.-J. Woo, Y.-S. Yuk, T.-H. Kong, S.-W. Honh, S.-T. Ryu, G.-H. Cho, A 40 mV transformer-reuse self-startup boost converter with MPPT control for thermoelectric energy harvesting, Proc. IEEE Int. Solid-State Circuits Conf., Dig. Tech. Papers, Feb, 19–23, 104–106, (2012).

[110] A. Shrivastava, N. E. Roberts, O. U. Khan, D. D. Wentzloff, B. H. Calhoun, A 10 mV-input boost converter with inductor peak current control and zero detection for thermoelectric and solar energy harvesting with 220 mV cold-start and $-$14.5 dBm, 915 MHz RF kick-start, IEEE J. Solid-State Circuits, 50(8), 1820–1832, (Aug 2015).

[111] A. Roy, A. Klinefelter, F. B. Yahya, X. Chen, L. P. Gonzalez-Guerrero, C. J. Lukas, D. A. Kamakshi, J. Boley, K. Craig, M. Faisal, S. Oh, N. E. Roberts, Y. Shakhsheer, A. Shrivastava, D. P. Vasudevan, D. D. Wentzloff, B. H. Calhoun, A 6.45 uW self-powered SoC with integrated energy-harvesting power management and ULP asymmetric radios for portable biomedical systems, IEEE Trans. Biomed. Circuits Syst., 9(6), 862–874, (2015).

[112] Z. Gao, Y. Zhou, J. Zhang, J. Foroughi, S. Peng, R. H. Baughman, Z. L. Wang, C. H. Wang, Advanced energy harvesters and energy storage for powering wearable and implantable medical devices, Adv. Mater., 36, 2404492, (2024).

[113] P. Simon, Y. Gogotsi, B. Dunn, Where do batteries end and supercapacitors begin?, Science, 343(6176), 1210–1211, (2014).

[114] S. Lage-Rivera, A. Ares-Pernas, M.-J. Abad, Last developments in polymers for wearable energy storage devices, Int. J. Energy Res., 46(8), 10475–10498, (2022).

[115] J. Ballard, M. Hubbard, S.-J. Jung, V. Rojas, R. Ung, J. Suh, M. S. Kim, J. Lee, J. M. Pierce, R. Venkatasubramanian, Nano-engineered thin-film thermoelectric materials enable practical solid-state refrigeration, Nat. Commun., 16, 4421, (2025).

[116] L. E. Osborn, R. Venkatasubramanian, M. Himmtann et al. Evoking natural thermal perceptions using a thin-film thermoelectric device with high cooling power density and speed, Nat. Biomed. Eng., 8, 1004–1017, (2024).

[117] -Q.-Q. Wang, K.-F. Liu, -Y.-Y. Su, X.-C. Liu, Q. Liu, S. Zhou, J. Liu, S.-Q. Xia, High thermoelectric performance and anisotropy studies of n-type Mg3Bi2-based single crystal, Acta Mater., 255, 119028, (2023).

[118] L.-D. Zhao, S.-H. Lo, Y. Zhang, H. Sun, G. Tan, C. Uher, C. Wolverton, V. P. Dravid, M. G. Kanatzidis, Ultralow thermal conductivity and high thermoelectric figure of merit in SnSe crystals, Nature, 508, 373, (2014).

[119] A. Nozariasbmarz, D. Vashaee, Effect of microwave processing and glass inclusions on thermoelectric properties of P-type bismuth antimony telluride alloys for wearable applications, Energies, 13, 4524, (2020).

[120] V. Leonov, R. J. M. Vullers, Wearable electronics self-powered by using human body heat: The state of the art and the perspective, J. Renew. Sustain. Energy, 1, 062701, (2009).

[121] V. Leonov, Thermoelectric energy harvesting of human body heat for wearable sensors, IEEE Sens. J., 13(6), 2284–2291, (2013).

[122] V. Leonov, Thermoelectric energy harvesting of human body heat for wearable sensors, IEEE Sens. J., 13(6), 2284–2291, (2013).

[123] M. Strasser, R. Aigner, C. Lauterbach, T. F. Sturm, M. Franosch, G. Wachutka, Micromachined CMOS thermoelectric generators as on-chip power supply, Sens. Actuators A Phys., 114, 362–370, (2004).

[124] K. T. Settaluri, H. Lo, R. J. Ram,Thin thermoelectric generator system for body energy harvesting, J. Electron. Mater., 41(6), 984–988, (2012).

[125] V. Leonov, Thermoelectric energy harvesting of human body heat for wearable sensors, IEEE Sens. J., 13(6), 2284–2291, (2013).

[126] S. P. Timoshenko, J. M. Gere, Theory of elastic stability ~mcgraw-hill, New York, (1961). A. Chajes, Principles of Structural Stability Theory ~Prentice-Hall, Englewood Cliffs, NJ, 1974.

[127] Q.-Y. Liu, X.-L. Shi, T.-Y. Cao, W.-Y. Chen, L. Li, Z.-G. Chen, Advances and challenges in inorganic bulk-based flexible thermoelectric devices, Prog. Mater. Sci., 150, 101420, (2025).

[128] M. Hyland, H. Hunter, J. Liu, E. Veety, D. Vashaee, Wearable thermoelectric generators for human body heat harvesting, Appl. Energy, 182, 518–524, (2016).

[129] A. Nozariasbmarz, H. Collins, K. Dsouza, M. H. Polash, M. Hosseini, M. Hyland, J. Liu, A. Malhotra, F. Mohaddes, F. M. Ortiz, V. P. Ramesh, Y. Sargolzaeiaval, N. Snouwaert, M. C. Özturk, D. Vashaee, Review of wearable thermoelectric energy harvesting: From body temperature to electronic systems, Appl. Energy, 258, 114069, (2020).
[130] J. Liu et al. Flexible thermoelectric generators compatible with commercial module fabrication processes international conference on thermoelectrics, ICT2017, California, USA, (Jul. 30th to Aug. 3rd 2017).
[131] F. Suarez, D. P. Parekh, C. Ladd, D. Vashaee, M. D. Dickey, M. C. Öztürk, Flexible thermoelectric generator using bulk legs and liquid metal interconnects for wearable electronics, Appl. Energy, 202, 736–745, (2017).
[132] S. H. Jeong et al. Stretchable thermoelectric generators metallized with liquid alloy, ACS Appl. Mater. Interfaces, 9(18), 15791–15797, (2017).
[133] Y. Sargolzaeiaval, V. Padmanabhan Ramesh, T. V. Neumann, V. Misra, D. Vashaee, M. D. Dickey, M. C. Öztürk, Flexible thermoelectric generators for body heat harvesting – Enhanced device performance using high thermal conductivity elastomer encapsulation on liquid metal interconnects, Appl. Energy, 262, 114370, (2020).
[134] W. Glatz et al. Bi2Te3-based flexible micro thermoelectric generator with optimized design, J. Microelectromech. Syst., 18(3), 763–772, (2009).
[135] T. N. Huu, T. N. Van, O. Takahito, Flexible thermoelectric power generator with Y-type structure using electrochemical deposition process, Appl. Energy, 210, 467–476, (2018).
[136] Z. Cao et al. Screen printed flexible Bi2Te3-Sb2Te3 based thermoelectric generator, in: Journal of physics: Conference series, Vol. 476, 1, IOP Publishing, (2013).
[137] S. J. Kim et al. High-performance flexible thermoelectric power generator using laser multiscanning lift-off process, ACS Nano, 10(12), 10851–10857, (2016).
[138] O. Bubnova, X. Crispin, Towards polymer-based organic thermoelectric generators, Energy Environ. Sci., 5, 9345–9362, (2012).
[139] Y. Wang, L. Yang, X.-L. Shi, X. Shi, L. Chen, M. S. Dargusch, J. Zou, Z.-G. Chen, Flexible thermoelectric materials and generators: Challenges and innovations, Adv. Mater., 31, 1807916, (2019).
[140] P. Yang et al. Wearable thermocells based on gel electrolytes for the utilization of body heat, Angew. Chem., 128(39), 12229–12232, (2016).
[141] C. A. Hewitt, A. B. Kaiser, S. Roth, M. Craps, R. Czerw, D. L. Carroll, Multilayered carbon nanotube/polymer composite based thermoelectric fabrics, Nano Lett., 12(3), 1307–1310, (2012).
[142] J. A. Lee et al. Woven-yarn thermoelectric textiles, Adv. Mater., 28(25), 5038–5044, (2016).
[143] T. Sun, B. Zhou, Q. Zheng, L. Wang, W. Jiang, G. J. Snyder, Stretchable fabric generates electric power from woven thermoelectric fibers, Nat. Commun., 11, 572, (2020).
[144] Y. Hou, Y. Yang, Z. Wang, Z. Li, X. Zhang, B. Bethers, R. Xiong, H. Guo, H. Yu, Whole fabric-assisted thermoelectric devices for wearable electronics, Adv. Sci., 9, 2103574, (2022).
[145] L. Zhang, S. Lin, T. Hua, B. Huang, S. Liu, X. Tao, Fiber-based thermoelectric generators: Materials, device structures, fabrication, characterization, and application, Adv. Energy Mater., 8, 1700524, (2018).
[146] K. Itoigawa, H. Ueno, M. Shiozaki, T. Toriyama, S. Sugiyama, Fabrication of flexible thermopile generator, J. Micromech. Microeng., 15(9), S233, (2005).
[147] M. Shiozaki, S. Sugiyama, N. Watanabe, H. Ueno, K. Itoigawa, Flexible thin-film bite thermopile for room temperature power generation, in: Micro electro mechanical systems, 2006. MEMS 2006 Istanbul. 19th IEEE international conference on, IEEE, 946–949, (2006).
[148] S. E. Jo, M. K. Kim, M. S. Kim, Y. J. Kim, Flexible thermoelectric generator for human body heat energy harvesting, Electron. Lett., 48(16), 1015–1017, (2012).

[149] A. R. M. Siddique et al. Thermal energy harvesting from the human body using flexible thermoelectric generator (FTEG) fabricated by a dispenser printing technique, Energy, 115, 1081–1091, (2016).
[150] A. Proto, J. Vondrak, M. Schmidt, J. Kubicek, O. M. Gorjani, J. Havlik, M. Penhaker, A flexible thermoelectric generator worn on the leg to harvest body heat energy and to recognize motor activities: A preliminary study, IEEE Access, 9, 20878–20892, (2021).
[151] W. Huo, Z. Xia, Y. Gao, R. Guo, X. Huang, Flexible thermoelectric devices with flexible heatsinks of phase-change materials and stretchable interconnectors of semi-liquid metals, ACS Appl. Mater. Interfaces, 15, 29330–29340, (2023).
[152] B. Iezzi et al. Printed, metallic thermoelectric generators integrated with pipe insulation for powering wireless sensors, Appl. Energy, 208, 758–765, (2017).
[153] Z. Lu, M. Layani, X. Zhao, L. P. Tan, T. Sun, S. Fan, Q. Yan, S. Magdassi, H. H. Hng, Fabrication of flexible thermoelectric thin film devices by inkjet printing, Small, 10(17), 3551–3554, (2014).
[154] J. Y. Oh et al. Chemically exfoliated transition metal dichalcogenide nanosheet-based wearable thermoelectric generators, Energy Environ. Sci., 9(5), 1696–1705, (2016).
[155] J. Weber, K. Potje-Kamloth, F. Haase, P. Detemple, F. Völklein, T. Doll, Coin-size coiled-up polymer foil thermoelectric power generator for wearable electronics, Sens. Actuators A Phys., 132(1), 325–330, (2006).
[156] L. Francioso, C. De Pascali, I. Farella, C. Martucci, P. Cretì, P. Siciliano et al. Flexible thermoelectric generator for ambient assisted living wearable biometric sensors, J. Power Sources, 196(6), 3239–3243, (2011).
[157] R. A. Kishore, A. Nozariasbmarz, B. Poudel, M. Sanghadasa, S. Priya, Ultra-high performance wearable thermoelectric coolers with less materials, Nat. Commun., 10(1), 1765, (2019).
[158] https://embrlabs.com/ (accessed on September 2025).
[159] https://www.dhamainnovations.com/ (accessed on September 2025).
[160] https://coolify.torraslife.com/ (accessed on September 2025).
[161] S. Krishnan. “Helmet with mechanism for cooling.” U.S. Patent Application No. 15/899,378, 2019.
[162] D. V. Chaplin, “Thermoelectric crash helmet cooling system with no mechanically moving components or fluids.” U.S. Patent Application No. 12/459,991, 2010.
[163] https://www.sony.com.hk/reonpocket/ (accessed on September 2025)
[164] N. P. D. Smyth, T. Hernandez, A. Johnson, Clinical experience with radioisotopic powered cardiac pacemakers, Henry Ford Hosp. Med. J., 22(3), 113–116, (1974).
[165] J. A. Paradiso, T. Starner, Energy scavenging for mobile and wireless electronics, IEEE Pervasive Comput., 4(1), 18–27, (2005).
[166] M. Ferrari, V. Ferrari, M. Guizzetti, D. Marioli, A. Taroni, Characterization of thermoelectric modules for powering autonomous sensors, Proc. IMTC Instrum. Meas. Technol. Conf., (2007).
[167] GreenTEG, Available online: https://www.greenteg.com/ (accessed on September 2025).
[168] https://about.nike.com/en/newsroom/releases/nike-and-hyperice-unveil-tech-enabled-boots-and-vest-designed-to-boost-athlete-warm-up-and-recovery (accessed on September 2025).
[169] Dual Temperature Mattress. Available online: https://sleep.me/ (accessed on September 2025).
[170] C. D. Kowal, J. R. Bertino, Possible benefits of hyperthermia to chemotherapy, Cancer Res., 39, 2285–2289, (1979).
[171] H. Xu, A. Martinez-Nicolas, W. D. Martinez-Avila, J. M. A. Alcantara, J. Corral-Perez, D. Jimenez-Pavon, F. M. Acosta, J. R. Ruiz, B. Martinez-Tellez, Impact of an intermittent and localized cooling intervention on skin temperature, sleep quality and energy expenditure in free-living, young, healthy adults, J. Therm. Biol., 97, 102875, (2021).
[172] N. Kubota, K. Okada, Y. Yamanaka, Circadian phase assessment of core body temperature using a wearable temperature sensor under the real world, Sleep Sci, 18(3), e246–e252, (2025).

Index

 | https://doi.org/10.1515/9781501514814-012